The Reliability Excellence Workbook:
From Ideas to Action

John L. Ross Jr.

Industrial Press, Inc.

Industrial Press, Inc.

1 Chestnut Street
South Norwalk, Connecticut 06854
Phone: 203-956-5593
Toll-Free in USA: 888-528-7852
Email: info@industrialpress.com

Author: John Ross
Title: The Reliability Excellence Workbook: From Ideas to Action
Library of Congress Control Number:

ISBN (print): 978-0-8311-3634-5
ISBN (ePUB): 978-0-8311-9491-8
ISBN (eMOBI): 978-0-8311-9493-2
ISBN (ePDF): 978-0-8311-9491-8

Editorial Director: Judy Bass
Copy Editor: Janice Gold
Compositor and Cover Designer: Janet Romano-Murray

books.industrialpress.com
ebooks.industrialpress.com

DEDICATION

For Bailey and Austin

TABLE OF CONTENTS

ACKNOWLEDGEMENTS

I always wondered what it would be like to write a book and add an acknowledgement section to recognize those individuals who were instrumental in getting me to a place where I thought I'd have enough to say to fill a book. My name did appear in the acknowledgement section of a best-selling military fiction book once. I felt very proud, although I felt I contributed little to the help the author craft a really good and suspenseful novel.

In my head, I imagined thanking my family and friends for the developmental years, and the support over the decades. Then I'd recognize those who I felt molded me into what I'd no doubt believed to be a level of expertise so that someone would actually pay to listen to me, or read what I had to say. In any case, I would end the acknowledgements in my mind by pointing out all those who had encouraged me to put pen to paper. These are the people that have worked to make sure that what I had to say made sense to at least someone other than me, and doing so with proper grammar and spelling

With that model in mind, I'd like to acknowledge my wonderful family. I grew up in the traditional nuclear family of the 1960's. My dad was an enlisted Army man, my mom was primarily a stay-at-home mom. I have an older sister, Regina, and a younger brother, Mike, both of whom currently reside in Oklahoma. We had a typical middle-class life, comfortable, yet adventurous. We camped a lot, and I always thought it was because it was cool. I now think it was because we didn't have much money. We never went without. My parents were great at being Mom and Dad.

My parents, John and Geraldine, have long been gone, but their influence on our family is constantly working its love and magic throughout the generations. My dad was without a doubt the smartest person I ever knew. In fact, it is possible he was the smartest person ever. He wrote a book once, an Army field manual on fighting Soviet tanks. He was fluent in seven languages, held four bachelor degrees, and we believed he was a spy. My dad could fly.

Despite my dad being the smartest person I knew, there was one other person that was just a little more capable. My mom. My Mom was the only member of our family who did not have a college degree, but she was the most capable person in the family. She taught three kids how to get dressed, make their beds, tie their shoes, cook, eat, and do dishes. She taught three kids how to balance a checkbook, work for a living, be respectful and kind to others, and love family above all else. My mom was the cornerstone of our family.

My sister and brother are the nicest and most genuine people you could ever hope to meet. We have a very close relationship to this day. I won the family lottery to be born between them. I am very proud of their accomplishments and the families that they have raised. I know that my family is the only reason I am able to write this today. There is nothing like family to give you the confidence to conquer anything.

When I was commissioned as a second lieutenant in the United State Air Force, my dad said, "I never met a second lieutenant I liked, so go out and change my mind." I always remembered that. I was fortunate to learn from some very capable professionals.

At my very first assignment, the senior NCO (Non-Commissioned Officer) that worked for me was SMSgt (Senior Master Sergeant) Robert Kalinyak. SMSgt Kalinyak had been in the service three years longer than I had been alive. He had twenty-six years in the Air Force, and I was twenty-three. SMSgt Kalinyak said to me on one of my first days, "Lieutenant Ross, I know you know a lot about aircraft, and engineering, but if you let me, sir, I'll teach you how to be a good officer." And you know what? He did. Thank you SMSgt Kalinyak. I think my dad would have liked you.

My first boss was Captain Steve Washington. I last knew Captain Washington as Lieutenant Colonel Washington. I worked for Captain Washington as a young Lieutenant. Captain Washington was a tremendous aircraft maintenance officer and a wonderful mentor to me and my fellow squadron officers. I like to think that it was under Captain Washington's supervision, that I was taught to learn from others, contribute to the conversation, make learned decisions, and give credit to the team when earned, and take the heat when things didn't go so well.

The last senior NCO I worked with directly, and I want to recognize him for his wonderful counsel, was SMSgt Steve Anderson. SMSgt Anderson looked like a recruiting poster for the Air Force. He was shaped like the giant letter 'V', and always had a perfectly pressed uniform. He wore taps on his shoes!

SMSgt Anderson once said to me, "Captain Ross, when a young airman tells me he can't do something, or won't do something. One of those, he said is a capability, and one of those is an attitude. I can tighten both of those up." And my goodness could he. SMSgt

Anderson was the finest NCO I ever worked with. We made a very good team. I hope he thought so as well.

I've never had a bad boss. I have been extremely fortunate. My very first civilian boss was Mike Crisp at Revere Ware Corporation in Clinton, IL. Mike was the Director of Operations, and I worked for him as the maintenance manager. Mike was very direct, "Don't come to me with a problem that you don't have a solution for. You won't like my solution. My solution, I'm home sleeping and you're here working." If you could make a compelling case to Mike, he would back you to the hilt. I really enjoyed working for Mike.

My boss, and still my friend, at Sheffield Steel was Gilbert Medina. Gilbert is perhaps one of the top ten steel makers in the world. I swear I saw him lick a chuck of slag one time and tell the melters what to add to get the chemistry right. If you have never been in the steel business, that last sentence will not make sense. Just know that Gilbert was one of the real talents in a very hard and aggressive business. I'm proud to call him my friend.

My boss at Griffin Wheel Company in Tulsa, OK, and later in Kansas City, KS was Eldon McDonald. Eldon was a tough one to crack, because he was a former maintenance manager himself. Under Eldon I really honed my project chops and we did some incredible work together. Eldon is a friend and was a great boss.

I've learned from my supervisors, but I never forgot to learn from those that worked with as well. I've already mentioned a few gentlemen from my military days, I'd like to acknowledge some other people that contributed to my understanding of maintenance and reliability from the other direction.

I have been very lucky to have some wonderfully talent maintenance managers working for me in my capacity as a plant engineer. I just want to acknowledge them and let them know that they had a big influence on me. C.L. Davis, Wayne Bishop, Claude Stout at Sheffield Steel, Brian Somolik at Griffin Wheel Company, and Steve Baumgartner at Farmland Foods.

I know we are going to be successful in the fight for greater reliability. How do I know this? Because we have people like these folks out there working on it every day.

Finally, I want to acknowledge those closest to the writing of this workbook, and those most influential in having me work in earnest.

I want to recognize the influence Dale Blann, Greg Folts, and Tracy Strawn have had on me as they are part of the Marshall

Institute ownership. I can honestly tell you no finer minds exist on the planet in regards to equipment reliability and service to others. I am remarkably lucky to be in the company of these gentlemen.

Many of my professional colleagues have contributed to the growth of my knowledge and experience with maintenance and reliability. Within this book, I've highlighted some of their very sage advice and thoughts on the matter. It seems prudent to mention those whose words are crafted into some of the points I make in this workbook. Thanks to Earl Porter, Steve Gowan, Mike Nelson and Mark Jolley for putting into words something that you can feel in your professional soul.

I want to thank Judy Bass for her guidance through the process of putting this book together. Thank you to Janice Gold for great editing, and Janet Romano-Murray for exceptional composition and cover art. The cover is awesome! I want to especially thank Mary Jo Richards for her work in constantly editing my work and providing an element of proper English to this workbook.

My great fortune to be born into a loving family, the luck of the draw in professional associations and growth, coupled with a gentle push over the finish line has contributed to what I hope you'll find to be a ready resource to help you move *from ideas to action*.

FOREWORD

ABOUT THE AUTHOR

The Beginning—Our Inherent Desire To Grow And Change*

"… it follows that any being, if it vary however slightly in any manner profitable to itself, under the complex and sometimes varying conditions of life, will have a better chance of surviving and thus be naturally selected."

—Charles Darwin, *Origin of the Species (p. 13)*

Continuous change is not continuous improvement. However, we are changing, of that there is no doubt. It is, after all, our destiny. The question really becomes, "Are we changing on purpose, for the better, or, are we changing by happenchance, and without prudence?" One form of change leads to sustained success and prosperity, the other to continual frustration and angst; to the point that we are likely to arrive at a place that is unrecognizable, and most certainly, not desirable.

It's a Marathon Not a Sprint

The rat-race is after all a race. The winner is not necessarily the swiftest, but the most innovative and adaptable. As you can see from Figures 1 and 2, the associate and the boss often have a very different idea about the route and time needed to win this race. For the business world, the *race* would be the corporate survival and the relevance of each process that is moved forward through continuous improvement.

---

* I thought about calling this section the "Introduction," but in reality, no one ever reads the introduction. So, I changed the title. I'm confident that no one reads footnotes either.

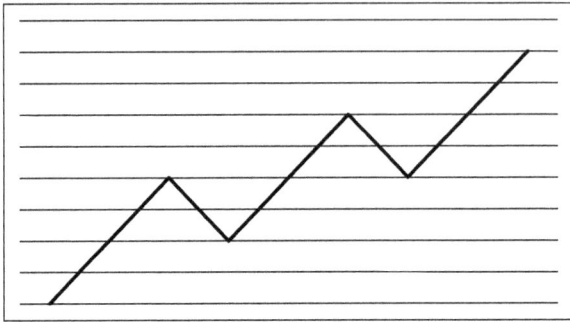

Figure 1: Actual continuous improvement

An organization's movement along the continuum of improvement efforts might very well look like Figure 1. There's something about the compression of time and the constant need for return on investment that really pales any sense of logic towards the slow, methodical approach needed for sustained success.

The contrast from the associate's thoughts are evident in the boss' timeline and desires reflected in Figure 2.

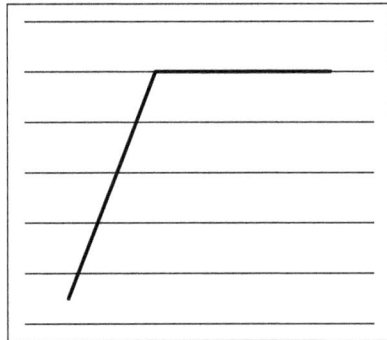

Figure 2: The boss's idea of continuous improvement

The very book you now hold represents a change from other books on the subject of maintenance and reliability. This particular volume is meant to provide a working document you can use continually to press the issue for prudent growth and yes, change.

This Workbook Can Change Your Professional Life

Throughout this book we will work together to understand the current state of your reliability processes, enable you and your allies to articulate the precise degree to which you are either winning or losing ground, and compel others into action, or support, in your natural progression forward. Each chapter will conclude with a synopsis of what we've learned together about your current situation, your vision, and we'll stitch together a patchwork of ideas to form a venerable and cohesive strategy. Together, we will be moving *from ideas to action.*

Darwin was quoted purposefully at the beginning of this section. He proved, through his studies, that incremental, slow growth will lead to meaningful change, for the better, or more directly, for survival. "*... If it vary however slightly in any manner...*" We must have a master plan for growth and success. To our knowledge, nature doesn't have a master plan But the slow, slight modifications to life, over thousands of generations, continues to afford the animal and plant kingdom an ability to survive and thrive in an ever changing environment "*... under the complex and sometimes varying conditions of life...*"

We don't have thousands of generations to wait, our need is more immediate. However, we can't ignore that for change to 'stick' it has to be change that is slow, intentional, and believed by the people to be *organic.* It turns out that we don't change culture. Instead, culture changes due to a new behavior exhibited over time. Make no mistake, change is coming. If it is not you that is personally positioned to develop the thought and vision, to lead the way forward, then who is it? If not now, for the positive change, then when?

The change, Darwin's thesis informs us, has to be profitable, or, in other words, have some value to those experiencing and living with the change. "*... profitable to itself.*" Profitability creates change that is sustainable and successful for the entities experiencing the change. Or, as Darwin tells us, those that change in this progressive manner will ultimately be "*... naturally selected.*"

This is a workbook, meaning that together we are going to work through this text to develop a master plan, or a strategy, to light the path forward from the status quo to a future state, or vision. This will be *your* path, these will be *your* words, and these will be *your* ideas and original thoughts. This workbook will simply aid you in developing the words you will need to compel others to join you in action to define and execute a well-reasoned reliability strategy. The work that we will do together in these pages will provide the 'money' argument needed in debate with those above you in the organizational structure, to secure support for all efforts going forward. The words and thoughts put together on the following pages will also provide the 'what's in it for me' answers required to gain support from those below you in the organizational structure. The plan that will form over the exercises, scenarios, and concepts in this volume we'll serve as a type of catalyst to engage you and your workforce to take charge and avoid just becoming victims of the change that is surely coming.

Roll up your sleeves. Let's get to work.

First Things First

This book is a workbook, so you're going to *work* to develop and record your thoughts. There will be some sharing of anecdotal stories, examples, and other experiences that you can add to your personal thoughts and ideas. The result will be a working document that delivers the mandate for our very survival, literally and figuratively.

Let's begin by establishing a common language.

George Bernard Shaw remarked once that England and the United States were said to be two countries "...divided by a common language." When it comes to the language of equipment performance and availability, there doesn't seem to be a common language at all. How can we achieve a common vision if we don't understand one another?

Here is your first task. This might seem elementary, but it is essential for establishing a framework of understanding. Let's agree in principle that we are not all defining the most basic terms the same. A common language is foundationally needed for a working relationship. In order for maintenance, engineering, production, purchasing, corporate leadership, and others to work and thrive together, understanding the most primary terms is required. Throughout this workbook, there will be shaded areas for you to "fill in the blank." Use the spaces below to record *your* definitions for the terms listed. We will be referring to these definitions throughout this workbook. Do not list textbook answers. This is a place to record your thoughts.

World-Class:

_____

_____

Reliability:

_____

_____

Uptime (or Downtime):

_____

_____

Optimize (or Optimal):

_____

_____

Effective:

_____

_____

Efficient:

_____

_____

Morale:

_____

_____

Maintainability:

_____

_____

Continue to add to the theorem that we don't speak a common language by asking your colleagues to define these terms and take time to explore what is almost certainly a difference in interpretation. We would suggest that you ask an operator, a production supervisor, a maintenance technician, a maintenance supervisor, a storeroom clerk, and a member of the front office, or plant staff.

By asking others to complete this most basic exercise, it will become apparent that, although we all want reliability, we just don't see it exactly the same way. That will be true for almost all the definitions.

It is important to get the 'street' definitions from your associates. Dictionary definitions don't work here, we need it straight from the heart.

Chapter Endings

Every chapter in this book will end with a review of the chapter's main points. These synopses will form segments of your overarching approach to the preservation of plant or facility equipment. Depending on the particular dynamics and circumstances, some of these chapter abstracts will fit, in the end, to outline a strategy for equipment maintenance and reliability at your location.

In the end, you will be able to review, edit, compile, and pull together the thoughts and ideas to form a very obtainable vision for the future, and make a conclusive argument, on any merit, for a purposeful drive forward.

What follows is an example, here in this introductory section, on what a chapter ending synopsis will look like.

Chapter Summary

Use the definitions and thoughts that you have collected in this chapter to form a basic maintenance philosophy, using words that your organization agrees best defines their objective.

What follows is the start of our overall maintenance and reliability strategy. It might be better argued as a philosophy, actually. Regardless, here starts the message we want to convey to our leadership. We'll continue to add to these chapter endings the themes to create a document we can actually use to move our reliability processes forward. Add your recently developed information where indicated. Please note that the chapter ending material also appears at the very end of this workbook. It is there, in the back of this workbook, that you will have a congruent flowing message of what you want your reliability program to look like, perform like, and deliver.

The developing philosophy:

Working in partnership, in a metered and measured manner, we intend to deliver world-class reliability efforts, meaning

_____

_____

And, attain and sustain the highest level of maintainability, meaning

_____

Our aim is to obtain the highest sustainable level of uptime, as measured by

_____

We are certain that an optimized approach to effective and efficient work, demonstrated as,

_____

will result in higher output and an increase in team morale, evidenced by

_____

Helpful Hints to Get the Most
From This Workbook

At the end of this workbook is a compilation of all the chapter summaries. This will provide you, the reader, a complete maintenance and reliability strategy, written with continuity and great purpose.

Work through these pages and engage your people and your partners to develop the philosophy and processes that work for you, in your environment.

Imagine presenting your associates and your boss with a thoughtful and pragmatic strategy to develop and execute the highest level of maintenance and reliability. That is the aim of this book.

This volume is also a perfect tool to educate your direct reports or your entire crew. Consider having each person in your organization fill in these blanks, and complete the exercises within this text. I expect this would advance the general knowledge of your decision makers, and enhance the engagement of your workforce.

The time is *now*, the person is *you*! Get ready to move *from ideas to action*.

ONE

Change and the Resistance To It

"In all cases, there is truth in resistance."

—Lisa Gundry and Laurie LaMantia,
Breakthrough Teams for Breakneck Times (p. 135)

Change is coming. In fact, it's happening all around us. "Resistance is futile," the Borg warned us in Star Trek. If so, why the hesitation? A comedian once asked, if the top boxer is the undisputed champion of the world, then "what's the fighting all about?" Those are both good points. Change is upon us so, really, what's the fighting all about?

There really is "truth in resistance." And, as it turns out, some of the resistance we are seeing is our own people telling us that we might be going about it in the wrong way. We might be forcing conformity to an issue that hasn't had time to gel, or grow organically within the organization. People don't resist change, per se, but as part of our human nature, we're just predestined to oppose most things that are unknown and uncertain to us. In a manner of speaking, it is this hard-wiring that has kept us alive this long.

Is it really 'change' that we are against?

It's a little ludicrous, and might be considered short-sighted, for leadership to assume people are resistant to change because no one likes to change. It's more likely that people are unwilling to go blindly along with change.

A somewhat controversial point was introduced in the opening segment of this book specifically that, "we don't change culture, culture changes due to a new behavior exhibited over time." It would not be unusual for someone in a leadership role to expressly

1

say, "We need to change the culture around here." Likewise, it is commonly noted, that many plant and division chiefs want their people to take ownership, meaning they 'own' their equipment and their processes. Given a little more thought, it might be unwise to just want culture change for the sake of change, and most leaders really aren't prepared for the employees to have true 'ownership.'

Getting To Work

We are going to work together, here in the opening segment to this chapter, to understand how we can influence, not change, but influence a shift in culture, and what real ownership means. Because we can't immediately change the culture, whether we want to or not, we should always look for and leverage the positive culture that exists in our companies right now.

List some characteristics of the current culture at your location:

_____

_____

_____

_____

Can we ever really just change the culture? All evidence says *no*. You may not even want to do so.

Here is a list of some personal attributes that might be part of today's culture at your location. Highlight those that apply to you and your associates:

- Come to work on time

- Produce a good quality product, or service

- Work in teams when asked

- Complete paperwork and documentation on time

- Keep work areas clean and organized at, or beyond the supervisor's expectation

- Stay over (overtime) when required

- Good citizens in the community

Look at the list that you've just highlighted, and this time ask the question, "What part of this current culture do we want to change?" As it turns out, the culture of your workforce may be just fine, and, any attempt to denude the workforce of their natural character through an un-natural culture shift, is the root cause to the resistance of change.

The Change Model

We don't change culture, culture changes over time. Figure 1-1 shows how culture is changed over time.

Of the connections linked in Figure 1-1, one illustrated link is between culture, and behavior exhibited over time. A different

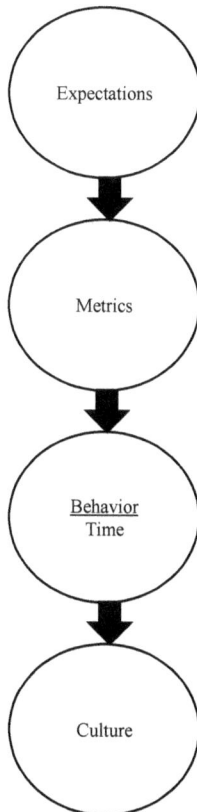

Figure 1-1: The change model

behavior over an unspecified, yet substantial timeframe, will result in a culture change. Sound like Darwin? The obvious follow-on question is, "What leads to a new (and acceptable) behavior over time?" The answer, as illustrated in Figure 1-1, is metrics. We use metrics to drive behavior.

Metrics

Example: Parents use their children's grades from school to drive a different, and hopefully better, behavior in class. When a child brings home a bad grade or an unacceptable report from school, a parent will likely instigate some level of discipline. A child might be grounded from a fun activity, or have some privilege removed until the behavior improves. The grade, or 'metric' is used to alter the behavior until a time where the threat (or consequence) is no longer needed.

Another example: A teenager learning to drive might receive a ticket for a moving or parking violation, or perhaps even get into a fender-bender in which they are at fault. The parent might take the teen's driving privileges away, or alter the freedom to drive until a period of time has passed and the teen has shown, through a 'better behavior,' that they are willing to comply with the rules of the road.

In both examples, the parent is using the metric of patterned improvement towards compliance to shift, alter, and drive the child towards conformance of an acceptable behavior. And, as we've discussed, all of this takes time.

The ultimate goal, in each example, is to develop young adults who will be successful in the culture of our world. We want our children to thrive in the culture of life without our constant oversight.

Where do metrics come from?

Expectations

As shown in Figure 1-1, metrics are derived from expectations. Expectations are established by the leader, and expectations come from the vision. Without a clear, concise, and understandable vision, there can be no set expectations. Expectations are metered by metrics and measures, which drive behavior over time, which eventually will lead to a thriving culture.

Expectations begets metrics, begets behavior over time, which begets culture change. Before all this begetting, something had to be begotten... a vision! Figure 1-2 shows our modified model.

The Status Quo

Figure 1-2 also establishes the status quo as the precursor to the vision. A colleague of mine once noted, "Even if you know where

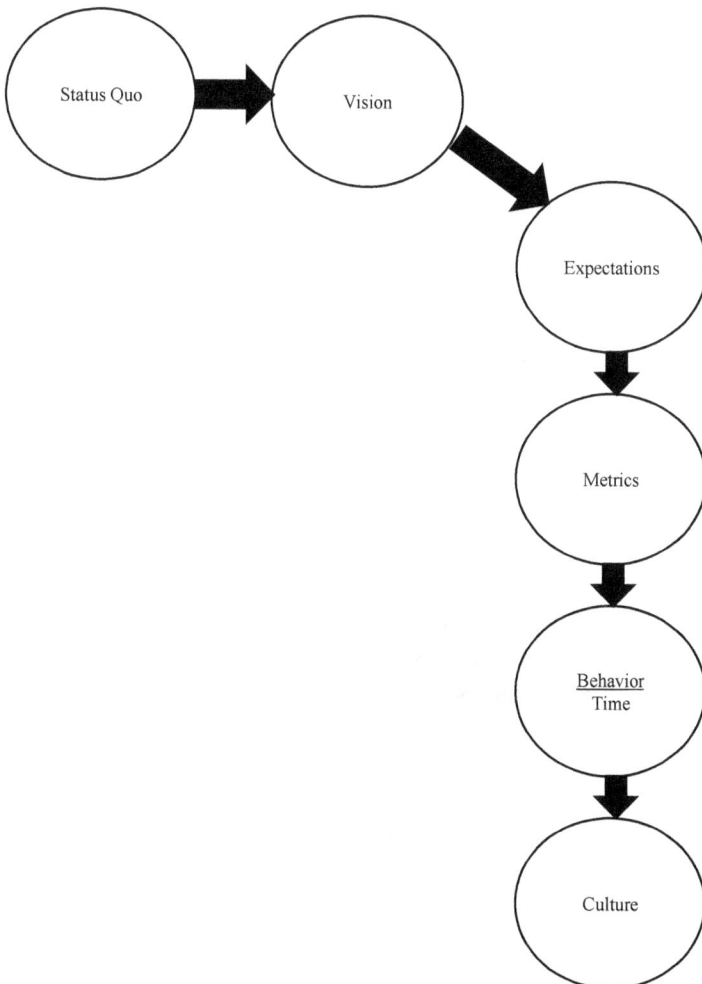

Figure 1-2: Expectations are drawn from the vision

you're going, you're still lost if you don't know where you are."
Defining for better or worse our current state, our status quo, is
essential for establishing the starting point of our path forward.
Planting a flag soundly in the ground of where you are now, clearly announces that *we own this*.

Are continuous change and continuous improvement synonymous?

Continuous Improvement (CI)

There is one final addition to the culture change model. Specifically, how does Continuous Improvement (CI) work with the new
order? Improvements must be made on top of the accomplished
and accepted improvements that preceded them. That's why it's
continuous improvement, meaning that some improvement had
to precede it. Continuous Improvement cannot be built upon a
change that was not an *improvement*.

The culture of the organization must arrive, en masse, at a
new cultural spirit before it can move on, en masse. Often it is
noted that a new leader for an organization, or a newly minted
leader within an organization, tries to hop-scotch over the entire
linkage from vision to culture and tries to shift the culture before
it is allowed to set up and establish that the new 'place' is indeed,
as Darwin stated, profitable to itself.

Figure 1-3 demonstrates the Continuous Improvement relationship with the change model.

From this line of deduction, we can conclude that meaningful
change, the kind that 'sticks' and can be built upon, must have
time to take hold, must follow from a vision that has been communicated, and must make sense to the masses. It is only from this
strong foothold that can we continue to grow and thrive through
continuous improvement.

Now it is time for you to review the level of serious change
taking place in your organization. People aren't so much resistant to change as they are unwilling to just go along with a change
that hasn't been proven to be profitable, agreeable, or even good
for them.

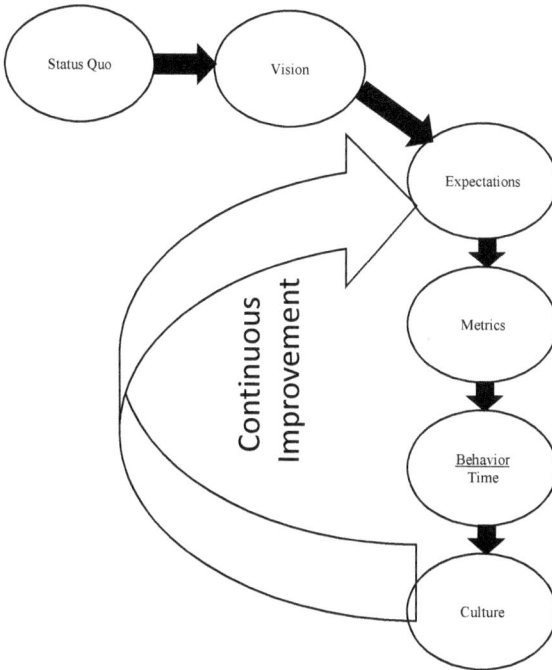

Figure 1-3: Continuous Improvement

Table 1-1 is an example of how a vision can be used as a link to the expectation we hope to achieve. From that expectation, we can link directly to a metric or measure, and then on to a behavior we are intending to modify. From that behavior there must be a profitable improvement seen by the workforce to entice them to move towards the vision.

Table 1-1: An example of how the vision links to the individual profit

Organizational Equipment Reliability Vision (as written, or believed to be)	Committed professionals applying top reliability practices today to secure tomorrow
Expectation #1 from that vision	Committed professionals
Metric/Measure from Expectation #1	Small group participation rate
Behavior to be modified, over time from Metric/Measure #1	Hourly associates drive improvement suggestions and idea creation
The 'profit' gained by the associates for exhibiting new behavior #1	Engagement, sense of importance and relevance for their processes and assets

Record your thoughts in Table 1-2. Start by documenting the vision and then expand through expectations, metrics, the behavior to be modified, and then the personal profit the associates will gain in this new place.

Table 1-2: The link between the vision and personal profit at your location

Organizational Equipment Reliability Vision (as written, or believed to be)	
Expectation #1 from that vision	
Expectation #2 from that vision	
Expectation #3 from that vision	
Metric/Measure from Expectation #1	
Metric/Measure from Expectation #2	
Metric/Measure from Expectation #3	
Behavior to be modified, over time from Metric/Measure #1	
Behavior to be modified, over time from Metric/Measure #2	
Behavior to be modified, over time from Metric/Measure #3	
The 'profit' gained by the associates for exhibiting new behavior #1	
The 'profit' gained by the associates for exhibiting new behavior #2	
The 'profit' gained by the associates for exhibiting new behavior #3	

If we can somehow miss this vision-culture connection, and we do, can we also manage to miss the ownership-authority connection?

Ownership

Having clearly established the link between vision and the presumed profit to the people, let's now shift this discussion on the reluctance to change to the managerial concept of *ownership*. Corporate and division leaders who state that, "I want our people to take greater responsibility, I want them to own the process" don't really mean it. How could they?

Today's leaders are very likely equating ownership to responsibility or accountability. In fact, their model might look similar to Figure 1-4.

Figure 1-4:
What the boss
thinks ownership
means

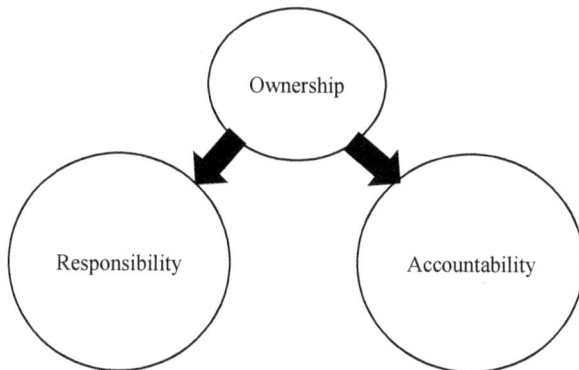

Ownership actually means Responsibility + Authority, as shown in Figure 1-5.

Figure 1-5:
What ownership
actually means

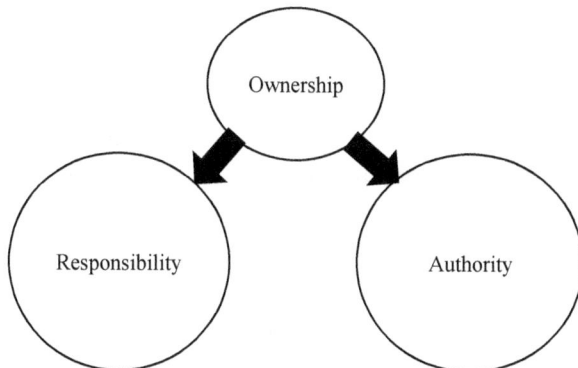

Very few leaders, it seems, are actually willing to concede authority to make decisions, instead they take action to generally maintain some level of absolute control. When the boss says they want their people to take ownership, they don't really mean it. Not really.

Example: An operator detects a small hydraulic drip on their machine. It isn't at a level of great concern, no machine operation is affected, and it's really just more of a nuisance because the operator has to clean a very small amount of fluid up each day off of their machine as a result. The operator has followed the instructions to enter a work request and has been waiting patiently for weeks to have some resolution. In fact, the operator has mentioned the leak to their supervisor, and to their supervisor's supervisor, and to a number of maintenance technicians that have walked by, or been on the equipment to perform some unrelated preventive or corrective work. The operator, justifiably, has grown weary of the additional work to keep their equipment clean, and is losing (if not already has lost) faith in the work order system.

At this point, in your plant or your facility, can the operator refuse to work on the machine until the small, inconsequential leak is repaired? If so, then that's real ownership.

By failing to tap into the strength of our workforce by facilitating their ownership in our facilities, we've essentially lost the corporate investment in what is called "intellectual capital." True ownership provides an internal mechanism to connect with our most valuable resource.

James Belasco and Ralph Stayer thought as much in *Flight of the Buffalo* (1993), noting that, "In today's intellectual capitalism world, the performers must be responsible for their own performance. The success or failure of the business must rest with the individuals who possess the critical capital" (p. 246).

Another example: An operator's day is likely to go better when their equipment runs as it is supposed to. One way that an operator can make certain their machine will run better is to ensure that the programmed Preventive Maintenance (PM) is accomplished correctly and on time. Can an operator in your plant refuse to run a machine that has not had its programmed maintenance completed? What if a production leader elects to run-through a

PM to get production out the door? Can the operator call foul and refuse to work? If so, that's real ownership.

We don't really want our associates to own the equipment or own the process, because we aren't ready, willing, and in most cases, able, to abdicate the authority necessary for *real* ownership. Power is a hard thing to abdicate

Are there any remaining questions as to why our members don't voluntarily march off in the direction in which we point? In many ways, any reluctance to change is a self-inflicted wound. As we will soon discover, we are all predisposed from influences in our lives, to move forward with caution and bias.

What we really lose in any hesitation to ease up on the reins and become a more flexible organization, is the loss of potential for greater things. In his work with Toyota, Jeffrey Liker noted, "The key difference between Taylorism [scientific management] and the Toyota Way is that the Toyota Way preaches that the worker is the most valuable resource—not just a pair of hands taking orders, but an analyst and problem solver. From this perspective, suddenly Toyota's bureaucratic, top-down system becomes the basis for flexibility and innovation" (2004, p.145).

If you'll allow a brief synonym linking ownership to empowerment, we might all be guilty of what Ron Moore recorded, "Empowerment appears to have come into popularity, and then waned, not because the concept isn't valid, but because it has been improperly applied" (p. 372).

Critical Thinking

To understand the approach we and our co-workers take to change, we have to understand how we approach the unknown. If change is coming, and it is, and if we will succeed when everyone moves towards the vision, and we will, why aren't more of us charging forward? It might not be a surprise that no two of us look at circumstances in exactly the same way. There are significant differences, of that we can be sure. However, on a micro-level, we are also capable of defining and selecting positive outcomes and feeling rewarded and justified in our own solutions.

It has always been an interesting phenomenon that two intelligent people can be so sure of different solutions to the same problem given the exact same facts. The prism they each use to interpret the information can often skew their reasoning. "One of the most important skills in critical thinking is that of evaluating information. This skill begins with the important recognition that information and fact, information and verification, are not the same thing." (Richard Paul and Linda Elder, p. 84)

This is a very brief discussion on Critical Thinking. Have fun with this segment and test your colleagues as well. You will discover that we all have more in common than we may have thought. We can solve complex problems, and achieve so much more by understanding the similarities and differences between us, and by leveraging the strengths in our abilities, and compensating among our team for the shortcomings we might expose.

The necessity to discuss critical thinking becomes evident when we confess that our decisions, our actions, and much of our current thinking is shrouded in a cloud of bias and we are often left to act on information that is not all together complete or clear. If we know, in the best case scenario, that our decisions typically lean towards our internal terms of reference, how much better are our decisions in the heat of battle; in an emergency or urgent situation? Paradigms and prejudices can, and often do, lead to wrong or less than our best decisions. In fact, we don't make our best decisions in a crisis; we make our most creative decisions. Therefore, it is imperative that we establish good practices through critical thinking, to ward off the known vices of our general problem solving skills.

In Table 1-3, check the box that best describes how you feel about the practice or process that is listed. For example, if you have a favorable opinion of the work order system, as a whole, check the box in the 'favorable' column. Have some of your colleagues do the same to garner a sense of the overall view of the processes listed.

If you are in a position of leadership in your company you undoubtedly have access to information that you may have called upon in order to make a judgment, favorable or otherwise. Your interpretation of the data itself may have been skewed if you were biased as to what level of performance was acceptable or unacceptable.

Table 1-3: Thoughts about current processes

Process		Favorable	Unfavorable
Work Order System			
Maintenance Planning			
Maintenance Scheduling			
Preventive Maintenance			
Predictive Maintenance			
Storeroom			
Operator Involvement			

If a production leader performed this exercise, they may have marked the practice of Operator Involvement as favorable; you might have marked it as unfavorable. If a maintenance technician had completed the exercise, they might have concluded that the Preventive Maintenance system was favorable; you might have marked it as unfavorable. Everyone was evaluating the same processes but from different angles and may have come out with different conclusions. That is both anticipated and reasonable.

Bias Isn't Always Bad

Our disposition to view situations based on our bias is not all together bad. What is critically important is for us to realize that we view situations and options through the prism of our paradigms; essentially through the hard wiring of our brains. This phenomenon keeps us from putting our hands on a stove's burner, even when we know it's not hot. We also have the mental training to look both ways before we cross the street and treat every gun as if it is loaded. Operating in a world without these synapses fused in our minds, would be chaotic and fraught with danger and error.

In business, and indeed in manufacturing, we have to apply a level of critical thinking to our reliability problems. This aspect gives us an ability to find the true nature of a fault and a viable solution, while keeping us from chasing symptoms. Mistakes in thinking aren't generally dangerous to man, but they can be fatal to the bottom line of a business.

You are going to supply a real world example of this fact. On the line below, list the name of one machine in your plant that suffers from chronic failure.

There is a good chance that the machine you listed is currently operating in your plant right now. It's not that your maintenance technicians don't know how to fix it; in fact they might have fixed it a dozen times this year alone. But they haven't really solved the root cause, not yet. Lack of willingness and creativity have never been the reason equipment fails, but repeating the same repairs and then continually limiting solutions to just 'inside the box' can keep us in a continuous hit-and-miss mindset. We can think and act our way to greater reliability.

Your machine may have been fixed a dozen times by technicians who employed their paradigm of what they thought was needed to fix the problem. This fire was further fueled by the operational supervisor's paradigm to get production out at all costs; manifested by a routine practice of running equipment until it fails and then asking the technician, "Why did it break, and when will it be fixed?" What's critical about the thinking in the machine example you provided above is that there was probably very little critical thinking involved.

There are three primary 'musts' to capitalizing on the concept of critical thinking relative to equipment reliability:

1. Understand that the nature of humans is to attempt to solve problems quickly, based on their paradigms; often with incomplete or too little information

2. Define and understand common critical thinking concepts used in plant maintenance activities so we know what to look for and what to avoid
3. Understand that when applying the elements and tools of Root Cause Analysis (RCA), we must be aware that our paradigms and prejudices can lead to false or unsubstantiated conclusions

Our response to situations is guided in many cases by human nature; these parameters have been established, over time, in our minds and thus have become our road map when solving problems.

Exercises

Here are a few simple warm up exercises to demonstrate how our paradigms and hard-wired thinking affect our response to even the simplest puzzles.

Exercise 1: Read the instructions below carefully and then answer the questions quickly.

Instructions: Read each question completely and then answer in the space provided. All those who previously served in the military must answer the questions.

Questions:
1. Complete the following number sequence:

$$2...4...6...8...\underline{\quad}...\underline{\quad}...\underline{\quad}$$

2. What is the probability that a quarter flipped in the air will land on heads?

$$\underline{\quad}/\underline{\quad}$$

What is the probability that a quarter flipped in the air one hundred times will land on heads one hundred times? (Hint: Don't confuse this with the 'likelihood' that it will land on heads one hundred times)

———/———

Exercise 2: Read the following paragraph carefully. There is a certain anomaly with the paragraph. Can you find the unusual characteristic?

> An angry outburst, such as saying "This is driving us crazy... " is a costly luxury for any consultant. Sponsors pay consultants for sound conduct and opinions—not angry outbursts. A part of my mind is naturally stingy, it loudly said, "No," to losing sponsors and softly said, "Log all angry outbursts." My log would un-shroud my angry outbursts. I was blaming things away from my own body and mind—things I can own and control -- for my thoughts. I took back my right to own my thoughts; which cut my outbursts by 50%. I did additional analysis of my log, I saw again and again angry thoughts snowball from small things to big things. I put a stop to this snowballing, which cut an additional 45% out of my angry outbursts. I thought I was looking good, but my stingy part's analysis hit with a thud: Although only 5%, 1 in 20, of my original angry outbursts now occur, my cost hadn't shrunk much. My conclusion is that any angry outburst is a high-cost pyroclastic display unworthy of a consultant.

Exercise 1 is actually two exercises in one. The first test is in the instructions, under the second point, "All those who previously served in the military must answer the questions." Many people taking the test will not perform the exercise (they will not answer the questions) because they read the second instruction bullet to mean that *only* those who have served in the military can/should answer the questions. Obviously, that is not what the instruction says. Read the instruction again and it should be clear that *everyone* should answer the questions and complete Exercise 1. Our brains have read something into the instruction that is not there.

The answers to the exercises are. 10...12...14; 50/50; and 50/50.

The answer to Exercise 2 is that the letter 'e' does not appear in the text at all; this is unusual for a paragraph of such length. The reader typically will not 'catch' this unusual characteristic because they read % as percent, and 45 as forty-five; which are clearly words that contain the letter 'e.'

At this time, it would be perfectly understandable for you to say, "so what?" What is the purpose of a discussion on critical thinking in a maintenance reliability text? Consider that, you, the reader, are of at least average intelligence; and that the exercises were very simple in nature. It is most likely that you did not respond correctly to all the exercise questions. Now, what if we were working on a production breakdown at two o'clock in the morning? The machine we are trouble-shooting is 20 feet off the ground, and it is one hundred degrees in the plant. How clear is our thinking? What if we are in a conference room, brainstorming the latest downtime causes? Do we enter the meeting with pre-conceived notions? Critical thinking is critical, but seldom understood or utilized. We need to understand the 'thinking' processes of those involved with the solution.

So, what is critical thinking?

Definitions

Critical thinking is a cognitive strategy that allows us to examine, debate, understand, evaluate, solve problems and make decisions based on sound reasoning and valid information. Critical thinking is paramount to finding root causes for equipment or process failures. *Root Cause Analysis* (RCA) will be discussed in a later chapter, but for now it is important to note that in RCA, much of our data collection and problem solving tools require working in groups, or among others, to mine for information and develop a pool of potential causes.

In the process of information and cause generation, people will participate and will comment as they are generally predisposed to do. As such, it is a valuable exercise to become aware of several factors and phenomena driving much of the input from your root cause analysis teams.

Critical thinkers have the ability to:

▪ Focus on a question

▪ Analyze arguments

▪ Ask and answer questions for clarification and/or challenge

Many, if not all, tools used in root cause analysis are clearly designed to be used with a group, or with an organized focus team. As a result, the tendency to drift off on a tangent can be overwhelming. The development of a problem statement is 'key' to arriving at a proper and successful conclusion. Expressing the elements of critical thinking will aid in *focusing on the question* (or problem).

Collectively, we must all *analyze arguments*—understanding that everyone is predisposed to approach problem solving with their own set of paradigms and prejudices will help in separating those arguments that are certainly a path to the solution, from those that would throw the process off course.

Ask and answer questions for clarification and/or challenge—ask for clarification when brainstorming, do not challenge suggested inputs at this time. Then, when utilizing root cause analysis tools during the analysis, challenge the thinking and the position of the participants. Understand that the solution is generally not always the most obvious (or simplest) choice. (Deference given to Occam's razor).

There are several pitfalls likely to be experienced during any discussion on problem solving or determining a proper course of action. These pitfalls are important to note so we are aware of their presence and can counter their often negative effects.

There is a certain proclivity for humans to confront the unknown with a somewhat defensive position. A few of the most likely pitfalls are listed in the following text to demonstrate some of the more commonly seen fallacies to clear and critical thinking. The importance of knowing about these pitfalls is to try and avoid them when possible. Study each of these examples and ask yourself if you've ever heard, or been guilty of, those types of 'misdirection' or tangential thinking in a group setting. Our inability to focus on the issues, and stay on task, often leads us down the wrong path.

Thus, we never really seem to solve the problems we constantly confront.

Here are some real examples. Take the time to highlight the ones that you have personally been guilty of or have experienced.

Examples

Phenomena and examples

Equivocation Fallacy

Using words that have double meaning—hinging on ambiguity.

> **Example:** Machine run time is critical to the company. Maintenance requires machines to be down to perform work. Therefore, maintenance is not critical to the company.

> **Example:** Having stocked spare parts is important to the maintenance of equipment. The storeroom tries to minimize the number of spare parts on-hand. Therefore, the storeroom is not important to the maintenance of equipment.

Begging-The-Question Fallacy

Presumption of a conclusion which is the question itself.

> **Example:**
> (Engineer) "I know this goes here, because I designed it."
> (Maintenance Manager) "How do you know?"
> (Engineer) "Because I designed it."

> **Example:**
> (Mechanic) "I've been a mechanic here for 25 years and that could not possibly be the reason the machine failed."
> (Newbie) "How do you know that?"
> (Mechanic) "Because, I've been a mechanic here for 25 years."

Ad Hominem Fallacy

Attacking the arguments of a position because of irrelevant personal circumstances.

> **Example:** That operator doesn't know how to operate their machine because he has such a bad attitude.

> **Example:** That electrician couldn't possibly have a suggestion on how to fix this problem because he is always so negative.

Appeal to Authority

Someone of great authority sways the argument, not because of their qualifications, but because of their position.

> **Example:**
> (You) "We'll fix that after we've addressed higher priorities."
> (Plant Manager) "Fix it now."

Red Herring Argument

Introducing a distracting issue of irrelevance into an argument.

> **Example:** We can never plan any maintenance activities because production never shuts down the machines.

> **Example:** We can't get anything fixed around here because the storeroom never has the right parts.

Fallacy of Conversion

If all A are B, then it stands to reason that all B are A.

> **Example:** All those in maintenance fix machines; therefore, all those who fix machines are in maintenance. This one is quite active in union environments.

Example: Operators break everything; therefore, all those who break things are operators.

Paradoxes

These are people, things, and ideas that exhibit an apparent contradictory nature. The example below illustrates a very popular mathematical paradox.

Exercise 3 (demonstrating the phenomenon of paradoxes): Read the scenario below carefully, this is a fairly well-known example.

Three men stop at a lodge for the evening. The innkeeper only has one room to rent. The cost is $30 dollars. Each man agrees to share the room and the cost. Each man gives the innkeeper $10.

After the men have retired to their room, the innkeeper decides that it was unfair to charge so much for three men to share one room. He decides to discount the room by $5. The innkeeper heads upstairs to return the $5. As the innkeeper climbs the stairs, he has second thoughts. He puts $2 of the $5 into his pocket and returns $3 to the men, resulting in a $1 discount for each man.

The problem:

- The room rented for $30

- Each man agreed to split the room and the cost

- Each man originally paid $10

- The innkeeper decided to discount the room by $5

- The innkeeper headed upstairs with $5 in his hand and $25 in the till ($30 original)

- The innkeeper had second thoughts and pocketed $2

- The innkeeper returned $3 to the men

- Each man received $1 as a discount

- Each man paid $10 and got $1 back

- Each man, therefore paid $9

The math: $9 x 3 men: = $27

And

The innkeeper put $2 in his pocket

Therefore: $27 + $2 = $29 dollars

So: Where is the other $1?

Circular or Tautological Reasoning

An inherent danger of only having an 'expert's' point of view. No other out-of-the box thinking can enter into the conversation.

> **Example:** Q: Why is Bob so smart on hydraulic systems?
> A: Because he is an expert.
> Q: How do you know he is an expert?
> A: Because he is so smart.

> **Example:** Which of the following four words does not belong with the other three?

>   A. Hand   C. Mind

>   B. Arm   D. Tool

If you exclude out-of-the-box, non-expert thinking, you might answer C since 'mind' is not a physical thing.

If you bring in some other people, you might be open to the suggestion that B is the answer, since it only has three letters and the other words have four.

Another person might suggest that D is an acceptable answer, because a tool is not associated with a living organism.

Dichotomous Variables

The conclusion is in the form of either/or.

> **Example:** Equipment is either running correctly, or it's not.

Note: There is a slight connection to this phenomenon and the *Kepner-Tregoe* process of 'Is or Is-Not.'

Cause-and-Effect

A causes B.

> **Example:** Not using proper torque values (cause) will result in bolts backing out (effect).

Note: This is a basic troubleshooting fundamental.

Naturalistic fallacy

'Is' vs. 'Should.'

> **Example:** Point: The machine *is* running slow.
> Counter Point: How fast *should* it be running?

Note: Always use empirical data when problem solving to avoid this situation.

Barnum Effect

> **Example:** That operator doesn't like to run his machine unless everything works perfectly. The Barnum Effect is the rhetorical question of "who does?"

Correlation vs. Cause-and-Effect

Correlation does not establish causation. You cannot prove cause-and-effect simply by establishing a correlation between events.

Example: A failing motor with bad rotor windings can trip a breaker. But a tripped breaker isn't always caused by a failing motor.

The correlation is the bad motor windings and the tripped breaker. The effect (tripped breaker) can come from many causes.

Continuity-Causation Error

Because two events occur at the same time, they are not necessarily cause-and-effect.

Example: Almost all superstitions are a parataxic mode of experience. If a ball player accidentally wears two different socks for a big game, and his team wins, it is likely that he feels a superstitious connection between the timing of wearing the mismatched socks and winning.

Example: We know that all components and all equipment are in a constant state of failure. When a machine breaks down, we commonly ask the operator what they were doing right before the failure. The continuity-causation error tells us that the operator's actions and the failure may only be connected in time, not cause.

Bi-Directional Causation

A to B (A can cause B or B can cause A).

Example: A misaligned sprocket can wear out a chain leading to the worn chain wearing out the sprocket.

Example: A worn sheave can wear out a V-belt leading to a V-belt causing premature wear to a sheave.

This is the classic "which came first, the chicken or the egg?" question.

Multiple Causation: Not Either/or But Both/And

Usually an effect occurs due to several causes.

This is why it is important to determine and address the root cause. A failure, due to many causes, can result in repeat occurrences. Also, when we troubleshoot by replacing parts, we often don't know what 'fixed it' or what combination of parts 'fixed it.'

> **Example:** A cracked machine frame could be caused by: loose mounting and/or excessive vibration and/or poor engineering and/or poor fabrication and/or poor installation.

Additional Examples

Intervention–Causation Fallacy

When the problem goes away, we didn't necessarily solve the cause.

> **Example:** We discover a bearing with a high heat signature. Our normal course of action might be to replace the bearing. After our action, the excessive heat is gone. Did we solve the cause? What caused the bearing to fail and to give off high heat readings?

This is the absolute fundamental element of Root Cause Analysis.

Deductive Reasoning

We begin with universal assumptions—this is theory driven.

> **Example:** Our theory is that loose chains lead to premature chain and sprocket wear. Loose belts lead to premature belt and sheave wear.

This is a fundamental basic in Preventive Maintenance.

Exercise 4 (demonstrating the phenomenon of deductive reasoning): Read the following scenario carefully. Answer based on your knowledge and experience.

Scenario: As you pull into the neighborhood tennis court complex, you observe a man in shorts, T-shirt, and tennis shoes exiting one of the court areas. He is holding a racquet, a can of tennis-balls, and he is covered in sweat.

What activity has this man just been involved with?

_____   _____
_____   _____
_____   _____
_____   _____

The truth is the man could have been involved with any number of activities; playing tennis would have been an obvious choice. He could have been shagging balls for other players, running around the courts for practice, just hitting balls, or any number of activities. We tend to want to think he was playing a game of tennis.

Inductive Reasoning

We begin with specific observations—this is data driven.

Example: Based on the chart in Figure 1-6, where is the next data point likely to fall?

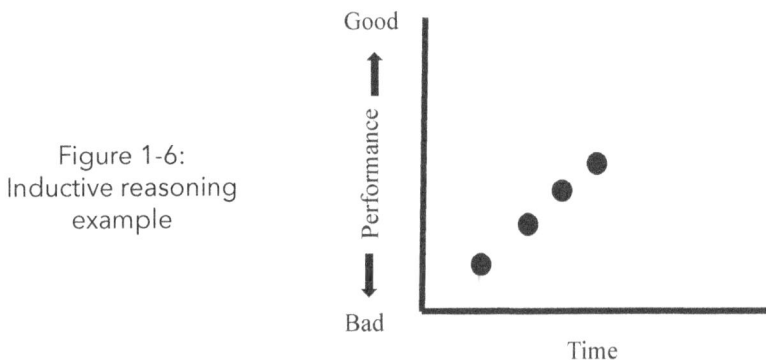

Figure 1-6:
Inductive reasoning
example

This is a fundamental basic in Predictive Maintenance.

Representative Heuristic and Representative Bias

This classic phenomenon plays out daily as maintenance technicians and engineers respond to situations based on what they know, or believe they know about causes. It is only an estimate that the probability of A caused B.

It mostly yields quick and often accurate results, but faults can occur because our information is inaccurate and incomplete.

A typical example is the occurrence of a technician replacing the same component over and over again, only to have that component "burn up" or fail in line as the others did.

"If I feel it, it must be true" Fallacy

A person's emotional comfort is not a valid manner to judge between what is true and what is false.

> **Example:** This is one explanation for the tendency of operators to adjust knobs and dials immediately when coming on shift and wondering how the off-going shift could have possibly "run this way."

Fallacy of reduction ad absurdum

If a contradiction can be drawn from an argument, the argument's assumptions are entirely false.

> **Example:** Misaligned couplings are presumed to give off an excessive heat signature. If a coupling does not give off an excessive heat signature, it is not misaligned.

Non-sequitur

The conclusion doesn't follow from the argument.

> **Example:** Brand X valve is inferior to Brand Y valve. Therefore, brand X valve failed because it is inferior.

> **Example:** More accidents occur at home. Of course they do, we spend more time at home than anywhere else. This is the law of averages.

"Strawman" Fallacy

A person attacks the weakest argument for a position, in spite of stronger arguments for the position.

> **Example:** Storeroom inventory is constantly attacked for the monetary value (weakest argument—in terms of reliability) when there are stronger arguments to support the inventory level (lead time, criticality).

There were twenty-four phenomena listed as examples of the pitfalls and dangers our problem solving or decision making efforts can face. It is most likely that as you read the passages above, you contributed many examples of your own on how things are discussed and decided at your plant or facility. The intent of this brief introduction to critical thinking was to simply make you aware that the converse to good and truly honest debate occurs frequently at our locations. The costs of entering the arena of problem solving with our time-learned responses is often a less than ideal ability to work together. We're predisposed to be biased.

How does this information help me now?

The term root-cause-analysis has been mentioned and it will be discussed in greater detail in later chapters; although at this point it might be reasonable to ask why a chapter on critical thinking was necessary at all, and certainly why so early in the text?

Those reading this material are coming into the context of equipment reliability from some established position. It might be from one that has been honed over many years and decades, or one that is at more of a novice level. Regardless, even your own perception of this text's material and ideas are subject to your mental filters. As an example, understand that when we discuss Key Performance Indicators (KPIs), you are automatically going to want to include some other KPIs that are not mentioned. Why, because that is what you perceive to be the correct thing to do; you are wired to think that. But is it truly part of the problem or the solution that we face? Or, simply how we have come to see good order and discipline in maintenance?

Discussions on critical thinking are important, and pardon the pun - critical, to ensure we understand that any conversation designed to find solutions, root out problems, or generate ideas is going to be fraught with predetermined reasoning and tangential forces. It is human nature, after all, to box and silo information into patterns that we can understand. This pretense does little to help find solutions, solve problems, or generate ideas. It usually takes us in the wrong direction, and leads to repeat disappointment. If we can understand, and give a name to the phenomenon constantly pulling at us, we can better negate its effects or eliminate them all together. We have to understand and resist its gravitational pull.

What Can We Do?

Now that we have a reasonable idea of some pitfalls to critical thinking and are somewhat convinced that we are all guilty of committing these and others in even the simplest problem solving or solution generating exercises, what are some good practices we should be exhibiting?

It would make sense to keep this discussion in the context of equipment reliability. Within that field we most likely will be working without updated or complete drawings and prints, and often without operating or maintenance manuals from the OEM (Original Equipment Manufacturer). This puts us in a predicament because there is a necessity to discuss absolute facts and conditions when engaged in robust dialog under the rules of critical thinking.

Often we are left with marked up prints, obsolete parts lists, and small notebooks of information to use in our efforts. It might be that we would find very little additional support material in our Computerized Maintenance Management System (CMMS). We really do enter problem solving situations armed with little or no information; just our gut feel and tribal knowledge.

The most vital aspect of problem solving or solution finding, is to first define the problem. Knowing what we do now about the pitfalls of critical thinking, we can sense right away that there is potential trouble with our 'team' or that we are about to travel a direction that will not net the best result.

Most problems are expressed as "the machine broke," or "the coupling failed," or "the operator keeps jamming the machine." Clearly, this first line of evidence will not put us closer to discovering the exact nature of the problem. The concept of problem definition applies easily to our topic of critical thinking as it does to root cause analysis as we'll discover later in the text.

In the context of addressing a perceived reluctance to change, it is critical thinking and the ideas we've recently been discussing that will help us to understand that outward hesitation isn't an attempt to derail the wheels of progress, but rather our life-learned approach to dealing with all things unknown. Understanding that can help guide us along the path to work as a team towards greater reliability.

Lean Maintenance

A great many have suffered needlessly through the advent and life of Lean Manufacturing. These are blunt and direct words that are meant to cast the shadow of truth on this methodology. Lean Manufacturing, as a movement, has become synonymous with 'downsizing.' Lean Manufacturing, after all, results in doing more with less, right? Of course that's not true. It's likely though, that everyone thinks that the purpose of Lean Manufacturing is to eliminate people from the process. When actually, with lean initiatives "… most of the progress comes because a large number of non-value-added steps are squeezed out. In the process, the value-added time is also reduced" (Liker, p. 31).

Imagine the zest to adopt a principle of Lean Maintenance. If Lean Manufacturing left a bad taste in our mouths, it's very likely that there will be enormous resistance to change in adopting its cousin—Lean Maintenance.

Ever heard of Lean Maintenance?

Lean Manufacturing's Influence on Lean Maintenance

So, what is Lean Maintenance and how is it similar and unlike Lean Manufacturing? First let's consider that in Lean Manufacturing, as we'll see in Lean Maintenance, the goal is to identify value added tasks (work) and non-value added tasks (work). The aim is to shore up and enhance the value added work, and to eliminate or minimize the non-value added work. As contrary as it sounds, some non-value added work is necessary, regulatory, or legally required; yet it is still non-value added.

What makes a task or work "value added?" There are three chief aspects for a task to be considered value added:

1. The work must physically change the 'thing' being worked on

2. The work must be done right, the first time

3. The customer must be willing to pay for the work (an assumption exists here that the customer knows of the work being done)

Exploring these points a bit further, there really is very little work accomplished, not valuable work anyway, that can be performed that doesn't physically change something. Paperwork, for example, is generally considered non-value added. After all, simply moving paper from one basket to the next, or to the filing cabinet has no intrinsic value. This is, of course, a very watered down example of paperwork. Some paperwork is very important, but is it really essential to what it takes to make something or to perform a service? Inspections are another form of work that doesn't change anything and therefore is most often considered non-value added.

Value added work requires that the work be performed right the first time. Effort should, therefore, be made to ensure that the first pass yield is 100%. It makes sense that work that has value is work that is done correctly without numerous attempts to get it right.

This dovetails nicely into the third point, which the customer has to be willing to pay for it. It's very unlikely that the customer would be willing to pay for something that doesn't physically change the item, and that isn't done right in the first place. The customer would generally not be willing to fund the scrap that is created in most processes and manufacturing.

The correlation between Lean Manufacturing and Lean Maintenance lies in the application of the three principles just described. Consider that maintenance work is value added if:

1. It physically changes the 'thing' that is being worked on

2. Is done right the first time

3. The customer (for our example,production) is willing to pay for it

Not all maintenance work physically changes anything. Is that work non-value added?

First, let's discuss maintenance processes, to include processes for the storeroom. It would be reasonable to assume that any process laid out for the effective and efficient operation of a maintenance department or a storeroom would be a process ripe with value. That is actually the result most often found when organizations take the time to write the process down. This is the genesis of the phrase, "it looks good on paper." However, it's more likely to be true that there are no processes written down, much less written down and followed. It would be more common to find undocumented processes in place that have evolved over time and are now simply "the way we've always done it."

It's hard enough to just develop a highly valuable and value added process. Imagine now having to encourage and cajole others to follow it. Would it be possible to arrive at such a level by accident and then sustain a level of excellence over time? It would be virtually impossible.

If done correctly, even a seemingly slight task can make a tremendous, value added impact.

Assessing Maintenance Processes—Practical Examples

Here is a classic example of the circular effects of maintenance, or better yet, poor maintenance.

Almost everyone coming across a copy of this book can give an anecdotal account of maintenance technicians removing items from the storeroom for legitimate work, but not recording any information, including the removal of the item, the issue of the part, the work order number, or, even their name. All the documented evidence indicates that the item just disappeared. The problem, in short, is that maintenance removes items from the storeroom, but doesn't write it down.

Small wonder really. The technician was in a hurry and the machine was down. The storeroom doesn't usually have anything and the technician was surprised the part was there at all. The technician might have the good intention of documenting the withdrawal at the end of the shift, but gets too busy to remember.

The process for removing items from the storeroom, especially those storerooms that don't have off-shift attendants, is to limit access to maintenance only, and train technicians how to record the items they took from the shelves. This often requires writing down the information; some companies have advanced to barcode scanners. There are other techniques as well.

Here is the entire replenishment process, of which that simple recording of the withdrawal is the lynchpin.

1. Technician enters the storeroom
2. Technician uses the CMMS or experience to locate the needed item
3. Technician removes the item from the shelf
4. Technician documents the removal with the item number, quantity, work order number, and name
5. Storeroom supervisor reviews the list of items removed during their absence
6. Storeroom supervisor deducts the quantity from the inventory count in the CMMS
7. Once, and if the inventory count for that item meets the reorder point...
8. Then the MRP (Material Requirements Planning) generates a purchase requisition for that item, ordering the reorder quantity
9. This reconciliation of "what hit the reorder point" happens every night, after midnight
10. Storeroom supervisor reviews the recommended order list prepared by the MRP
11. Once approved, the purchase requisition is essentially greenlighted for a purchase order

12. Based on the information in the Item Master Data, the item is reordered with the reorder quantity
13. The vendor receives the purchase order
14. The vendor may or may not have the item at their location, and may have to order it from the factory or a distribution center
15. The item may go to the vendor for shipment to you, or might be drop shipped directly
16. The item arrives at your receiving dock
17. The receiving dock is not in the storeroom, so there is a delay in the item getting to the storeroom
18. Once received in the storeroom, the item's BOL (Bill Of Lading) is reconciled with the purchase order and confirmation that everything is correct
19. The storeroom clerk puts the item back on the shelf
20. The storeroom supervisor adds inventory to the count in the CMMS

After the third bullet point, none of what was listed will happen until that technician records that withdrawal. The process has to demonstrate the value to the customer; which in this case is the technicians themselves (irony). And these steps need to be done in a timely manner and correctly, each time. The physical change is that the item is physically gone.

A side note, yet very interesting aspect to the replenishment process described above, is all the while, the clock is ticking on a stocked item that is at or below its minimum level. Thus the urgency to perform all the steps correctly, the first time, without delay.

On the subject of preventive maintenance; specifically after mentioning that most inspections are non-value added. Consider that production will always pay for superior performance. That is to say that they will provide equipment for maintenance that actually results in a favorable outcome.

Given that there are many types of preventive maintenance, and in this example we are only considering the 'inspection' type preventive maintenance (PM), let's agree that an inspection, only, does not physically change the item and therefore fails in the very first element of what makes a tasks or work value added.

Here is a very vivid example, it's written to the extreme to demonstrate the absurdity of much of our practices. Some metrics suggest that upwards of 60% of our available maintenance hours should be spent in proactive work, specifically preventive and predictive. There are numerous types of PM (discussed in a later chapter), but for this instance, let's assume that 50% of that 60% is dedicated to only inspection type PMs.

Scenario: Your maintenance department has twenty-five technicians, all scheduled for forty hours of work per week. This week, you are losing ten full man-days to vacations, meetings, and other assigned, non-maintenance, work.

$$
\begin{array}{r}
25 \text{ technicians} \\
\times \quad \underline{40 \text{ hours}} \\
1{,}000 \text{ man-hours}
\end{array}
$$

$$
\begin{array}{r}
1{,}000 \text{ man-hours} \\
- \quad \underline{80 \text{ (10 full man-days)}} \\
920 \text{ man-hours available}
\end{array}
$$

$$
\begin{array}{r}
920 \text{ man-hours available} \\
\times \quad \underline{60\% \text{ (percentage dedicated to PMs)}} \\
552 \text{ man-hours dedicated to PMs}
\end{array}
$$

$$
\begin{array}{r}
552 \text{ man-hours dedicated to PMs} \\
\times \quad \underline{50\% \text{ (percentage dedicated to inspection type PMs)}} \\
276 \text{ man-hours dedicated solely to inspection type PMs}
\end{array}
$$

Around 30% of the overall available man-hours for the entire week will be dedicated to inspection only type PMs. This is work we've just defined as non-value added for the sole reason that nothing is being physically changed.

Note that the other two characteristics of value added work are present (hopefully) in our inspection type PMs: done right the first time, and the customer is willing to pay for it. Or, are they?

Where does that leave us?

1. Inspection type PMs, which by definition are non-value added because nothing is physically changed

2. Done right the first time, yet many of our inspection PMs result in machines that won't start, or failures of the very components we just PM'd (leaving Plant Managers to ask, "Who PM'd that last?")

3. The customer is willing to pay for PMs that stick, that is to say, the PM results in a component that lasts. More on this in the section on preventive maintenance.

Some non-value added work is necessary, that much was explained at the beginning of this section. There are two types of non-value added work: Type 1 and Type 2. Type 1 non-value added work is necessary, required either by regulation, law or policy. Type 2 is simply waste-truly non-value added.

We have to work to ensure our PM inspections merit the time taken to perform them, and the time surrendered on production or service equipment.

Whether Lean Manufacturing or Lean Maintenance, our attempt should always be to look for evidence of waste and work to eliminate the wasteful tasks or at least minimize the effort taken to perform the task.

Waste

Waste affects our ability to *lean out* our maintenance operations.

Most, if not all, are aware of the classic seven forms of waste as communicated in Jeffrey Liker's work, *The Toyota Way* (2004, p. 28).

1. Over Production
2. Waiting
3. Transporting
4. Over Processing
5. Inventory
6. Excess Motion
7. Defects

An easy acronym to remember, shared by a colleague of mine, is: TIMWOOD

- Transporting
- Inventory
- Motion (Excess)
- Waiting
- Over Production
- Over Processing
- Defects

As it relates to Lean Maintenance, much can be gained by out-right elimination of waste, or at least minimizing the waste built into maintenance tasks and work. This has to be done in the spirit of delivering world-class reliability to the production effort, whether in the manufacturing or service industries.

Use the space below to record examples of these forms of waste in your maintenance organization:

Transportation: _____

Inventory: _____

Motion: _____

Waiting: _____

Over Production: _____

Over Processing: _____

Defects: _____

If you find that you're having trouble completing this exercise, ask someone else for their input.

Six Sigma

It would be reasonable to ask why there is a section about Six Sigma in a book about equipment reliability. Fair enough. It might be nice to think that if W. Edward Deming were alive today, he'd be an expert at Six Sigma given his statistical background. In many ways, Six Sigma and reliability are complimentary methodologies. They certainly are influential to one another.

Any organization in the world that is purposefully engaged in some sort of continuous improvement effort is working to increase throughput, improve quality, or increase reliability. Maybe all three.

Connecting Six Sigma to Lean to Total Productive Maintenance (TPM)

Companies desiring to improve throughput in their organizations often consider using the methodology of Lean Manufacturing. Lean Manufacturing provides the tools and techniques to understand how work is 'pulled' through the plant or organization. Lean addresses the seven forms of waste identified earlier, and explores ways in which more work can be done, with less effort. A popular phrase coined by an associate is, "get the most-est for the least-est."

Companies hoping to improve the quality of their product or service are likewise bent to adopt a leading methodology for quality improvement—Six Sigma. Where Lean Manufacturing addresses value added vs. non-value added, Six Sigma seeks to understand and control (minimize or eliminate) the variations in a process or service.

Interestingly, neither Lean nor Six Sigma addresses equipment reliability at the root level. However, all can agree that the unreliable nature of equipment can affect the ability to produce the product (throughput), and also, with equipment breaking down or straying off a center point, the quality of the resulting product is sometimes compromised. Because of that, the third element of continuous improvement, *equipment reliability*, is necessary.

The methodology most recognized as the foundation for equipment reliability improvement around the globe is Total Productive Maintenance (TPM). Regardless of the reliability philosophy you might follow, all forms have roots in Seiichi Nakajima's work on the subject, and subsequent book of the same name. Figure 1-7 shows the three most prominent methodologies for continuous improvement.

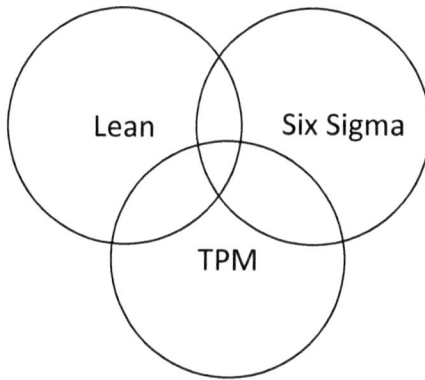

Figure 1-7: Three common Continuous Improvement methodologies

The matter of Lean Maintenance was addressed in the previous section, and there is little cause to detail Total Productive Maintenance, as this author would do little more that relay what has already been accepted as common practice on the subject. What is necessary, here, is to tie in the idea of Six Sigma, or better yet, variation elimination, in the maintenance processes and practices.

Several years ago, while working at a major manufacturing facility in the Northwest, a union lead-man made a very valid point when countering a suggestion that he should spell out the procedure to perform a certain PM task. "I don't have to tell my guys how to check a bearing, they know how to check a bearing," was his argument.

Nick had thirty-two mechanics that he was responsible for, and truly, each person had the capacity and experience to "check the bearing." The response to his point was classic Six Sigma, "I know your guys know how to check a bearing, but you need for them to check the bearing like YOU tell them to check the bearing."

If variation in the execution of maintenance processes and practices are to be minimized if not all-out eliminated, it makes sense that the acceptable process be written down, followed, and audited to ensure compliance.

Six Sigma is not just a tool to use to ensure a product matches specifications and the customer's exacting standards. It is also a tool to use to make sure that services are on point and meet the criteria for success.

A further point suggested to Nick was, "Today, you and your guys can establish the best way in the world to 'check' a bearing. Write that down, train everyone to that standard, and then execute."

The potential for variants in maintenance work are common place.

Variances in How We Execute Our Work

Opportunities to introduce variation in the field of maintenance exist anywhere information is exchanged, or 'work' is executed. It also lies in the spare parts that are in stock, or ordered for a repair activity. In the Air Force, this would be considered a target rich environment.

Procedures and processes that are not written down cannot be followed. Associates cannot be audited against policies that are not recorded. If employees are not trained to the correct standard, they cannot be expected to follow a correct standard.

Establishing the correct (as it is known at the time) process, and training others to understand and follow that process is critical to arriving at an expected and successful outcome. As that much is true, Six Sigma becomes the avenue most travelled to net minimal variation in the maintenance process.

Table 1-4 has an area for you to record some steps most commonly taken at your location to 'check a bearing.' After listing a few steps, record the potential for variation in that process. If your organization does not have such an inspection, use another example that would be common to your industry.

Table 1-4: Activity and a potential variant

Step	Activity	Potential Variant
Example	Check for side play	None, slight, a little, a lot, etc.
1		
2		
3		

Do Something Awesome for Six Months

This is meant to be a fun chapter ending session designed to encourage the reader to just do something awesome for six months.

Years ago, a very learned reliability professional presented a position on measuring that goes against some well-established thought on the subject of metrics. He most likely wasn't intending this to be a standalone idea, but it really makes sense in the abstract.

The expert said, "It's not the number you're after, it's the delta." Meaning, when you measure a process, initially, don't get upset or down-trodden by the number itself. After all, it's your initial measure, it's supposed to be low. What you're really after is the change, the delta, from then-to-now. Measure where you are, he says, then, "Do something awesome for six months, and see where you are then." That's the delta.

Consider the theme established at the start of this text, the slow incremental change that leads to lasting success and survival. Measure or evaluate something, make a slight change, and do that awesome thing for six months and measure again.

Theories indicate that anywhere from twenty-one to sixty-six days is required to form a habit; some suspect the same amount of time might be required to break a habit. In regards to what has been discussed, if a slight change takes you in the wrong direction it won't take twenty-one days to notice.

> *"We all want progress, but if you're on the wrong road,*
> *progress means doing an about-turn*
> *and walking back to the right road; in that case,*
> *the man who turns back soonest is the most progressive."*
> —C.S. Lewis

Chapter Summary

Use the ideas generated in this chapter to continue defining your approach to equipment reliability. We now know that reluctance to change isn't so much a criticism of the associates, but rather the result of the overall ineffectiveness to paint a viable vision

and establish expectations. Further, along that path, employees
aren't equipped with the authority to take real ownership. All
this is fraught with peril when we honestly account for the different biases people bring to the table when working together for a
common goal.

Our developing philosophy (uses notes from Table 1-2):

Our reliability vision is _____
_____
_____
_____

As such, we have expectations to deliver on these three specific
points (uses notes from Table 1-2):

1. _____
_____; measured by _____
_____; resulting in _____
_____.

2. _____
_____; measured by _____
_____; resulting in _____
_____.

3. _____
_____; measured by _____
_____; resulting in _____
_____.

Furthermore, we believe that real ownership is derived by
connecting responsibility with authority. Maintenance, production, engineering and purchasing will work together to ensure that
those associates assigned the responsibility and accountability for
a machine's reliability, or a process, will have a clearly defined
level of authority. Additionally, the path to reach a higher level of
authority will be unambiguously stated.

Improvement does not come from reliability increases alone. Additionally, we will work together to address waste in the area of maintenance, specifically:

Transportation: _____

Inventory: _____

Motion: _____

Waiting: _____

Over Production: _____

Over Processing: _____

Defects: _____

And, we will work as a team to minimize or eliminate areas that can contribute to undesired variation in the service and products of the reliability efforts. Variations such as:

- Ambiguous task instructions

- Lack of specificity in PM steps

- Open ended maintenance time allotment

- Storage conditions of storeroom items

TWO

The Business of Reliability

"If one listens to the faintest but constant suggestions of his genius,
which are certainly true, he sees not to what extremes, or even insanity,
it may lead him; and yet that way, as he grows more resolute and faithful,
his road lies. The faintest assured objection which one healthy man feels will
at length prevail over the arguments and customs of mankind.
No man ever followed his genius till it misled him."

—Henry David Thoreau, *Walden (p.165)*

Equipment reliability is a business. Maintenance is a business. One of the tactical errors we've made over time is not realizing this. In fact, maintenance professionals have been in error by not espousing the virtue of equipment reliability as a core value driver of the enterprise, regardless of industry. As a greater sin, we haven't had the vehicle to express our thoughts on how to manage the business we find ourselves engaged in.

What we are to learn and take to heart in Thoreau's passage is how to define our own success. And beyond that, how to find purpose and do what we love. How better are we able to do this, in an industrial work setting, than to be a part of the design of the vision and mission of the *thing* we are doing? It's impossible to be a part of the whole without being a part at all. Our genius will not mislead us.

We have ideas and enthusiasm about the things we are excited about. We can leverage this for the good of the business and grow personally and professionally as a result. We must be engaged with our enthusiasm.

Everyone in the company is, or should be, engaged in equipment reliability. In fact, this is the central premise of Total Productive Maintenance. We are *all* responsible for equipment reliability. That's right. Everyone in your organization should be able to, with a little thought, describe how the execution of their roles and responsibilities contributes to the capital equipment running correctly and reliably.

We are going to demonstrate this thesis together. Name one critical, or very important piece of capital equipment at your location. List it here:

_____

_____

_____

_____

The Ties That Bind

If this piece of equipment is truly important, everyone in the organization, to some degree, owes their very paycheck to the effective and efficient use of that equipment, along with its longevity. After all, the manufacturing and service industries are capital intensive. It takes a lot of money (capital) to run a business. Your equipment's reliability essentially pays your home mortgage, your light bill, buys your groceries, and in a sense, determines where your children will go to college.

For example: Mechanics, executing good maintenance practices, help to ensure reliable equipment.

Another example: Accounts Payable personnel, adhering to negotiated payment schedules, facilitate good relationships with vendors and suppliers for products and services, and therefore help to ensure reliable equipment.

Table 2-1 lists common roles that might exist in any organization. Make a connection between the role identified and how that per-

Table 2-1: Roles and their connection to reliability

Role	Connection to Equipment Reliability
HR Specialist	
Safety Manager	
IT Manager	
Storeroom Clerk	

son contributes to equipment reliability. Use the examples shown previously as a guide.

It is the execution of our roles and responsibilities towards asset reliability that oils the financial engines of our companies.

Only Assets Make Income

Robert Kiyosaki's best seller, *Rich Dad, Poor Dad*, is a helpful guide used to educate people on how the rich think about money. The premise is to understand and mimic those attitudes about money and wealth, in hopes of attaining the same sense of growth and prosperity in your own life.

Borrowing a diagram from Kiyosaki's book is perhaps the most effective way to demonstrate that equipment reliability is critical to business success. More to the point, having the correct maintenance processes and practices, and then executing the same, are true assets to any enterprise.

Kiyosaki addresses four elements:

1. Income

2. Expenses

3. Assets

4. Liabilities

Profit and Loss, or P&L, is comprised of income and expenses. The P&L, or income statement, "shows whether your business is profitable." (R. Abrams, *Business Plan in a Day*, p. 114) The Balance Sheet is made up of the assets and the liabilities and "shows the value of your company." (Abrams, p. 114)

Assets, as Kiyosaki instructs, make income. This is important, because in the world of money, it is *only* the assets that make income. Income comes from the manipulation of assets. In money terms, this could be equity in a home, stocks that pay dividends, or other money generating assets. A car, for example, is not an asset unless it is being used to make money. Consider this point when reviewing Figure 2-1 on the following page.

Income from assets is used to pay expenses. In our personal lives, that would be the light bill, water bill, etc. In a business setting, expenses would include those bills, as well as the payroll. Figure 2-2 shows how we use income to pay expenses.

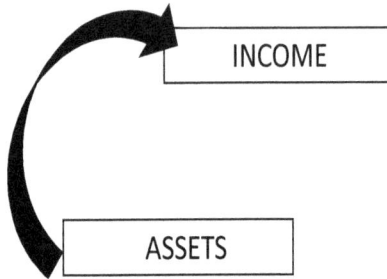

Figure 2-1: Assets 'make' income

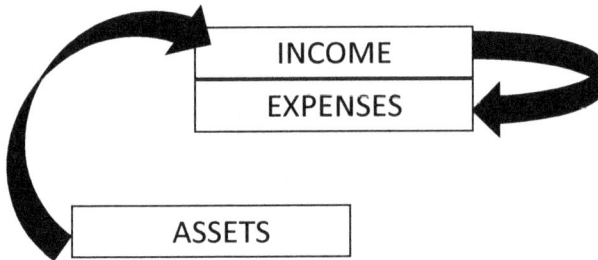

Figure 2-2: Income is used to pay expenses

Only after expenses are satisfied are the liabilities serviced. For an individual, these liabilities might include monthly mortgage payments, interest on credit cards, taxes, and interest on loans or liens. For a business, these same categories need to be addressed. The liabilities in Figure 2-3 are satisfied after the expenses are paid.

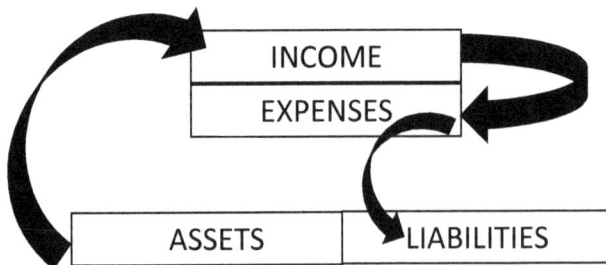

Figure 2-3: Liabilities are addressed after expenses are paid

The result of this, illustrated in Figure 2-4, is profit.

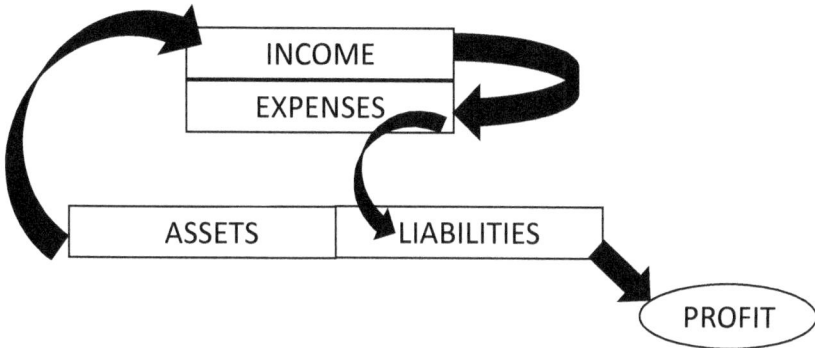

Figure 2-4: Profit is the goal

The goal is *profit*. The goal for an individual is to end the month with some money left in the checking account. For an organization, the desire is to end the fiscal year with a profit that can be reinvested (as an asset) back into the company to make more income and therefore more profit.

Forget for a minute that the CEO or owners of the company might buy a bigger yacht and take a European vacation with some of the profit. The true profit is the remainder of the income after deducting expenses and liabilities. A company will reinvest the profit back into the company in order to make more profit. It is likely that investors or company owners continue to put money back into the organization because they feel there can be more gain for themselves in the business than on the stock exchange or in bank products.

Everyone in the organization has, as a responsibility for their role, an understanding that their work should be profitable to the company. On the subject of equipment reliability, everyone has either a tacit or implicit responsibility to make sure the operation runs smoothly and reliably. When capital equipment is involved, everyone is responsible for equipment reliability.

Maintaining reliability is a business, short and simple.

Examples

Figure 2-5 shows a few examples of assets, as well as an example of an expense and a liability.

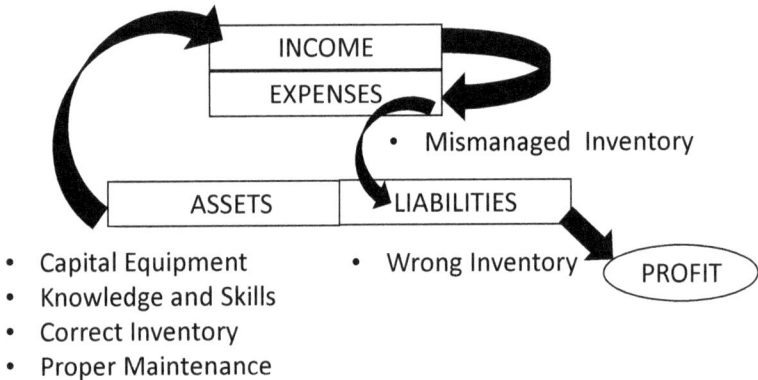

```
                    ┌─────────────────┐
                    │     INCOME      │
                    ├─────────────────┤
                    │    EXPENSES     │
                    └─────────────────┘
                              • Mismanaged Inventory

        ┌─────────────┐ ┌─────────────────┐
        │   ASSETS    │ │   LIABILITIES   │
        └─────────────┘ └─────────────────┘
                                              ┌─────────┐
 • Capital Equipment      • Wrong Inventory  │ PROFIT  │
 • Knowledge and Skills                       └─────────┘
 • Correct Inventory
 • Proper Maintenance
```

Figure 2-5: Category examples

Are your maintenance practices an asset? If your plant spent $10,000,000 on maintenance last year, what do they have to show for it? You can't count production's result because that was production's output, not yours. What was the output of the maintenance organization? Was it worth $10 million? If not, then at your location maintenance might not be an asset after all. It's very likely considered an expense, a necessary evil. Or worse, a liability.

What has been the approach to entities that contribute nothing to the bottom line, yet cost money in your organization? In the old days, those agencies would be minimized, deleted, sidelined, cut, or eliminated.

A maintenance organization is nothing more than people executing processes and practices. If those processes and practices cannot be shown to be 'assets' that make the company some income, then perhaps the maintenance department should be eliminated. That's business.

A company that places a corporate and strategic value on equipment reliability is a company that most likely performs many aspects of its business at an exceptional level.

Have you ever really considered what it looks like to be in a high performing company?

A Company That 'Gets It'

Figure 2-6 shows an example of the kind of activity that a high performing company might be engaged in.

Look at this photo and ask yourself, "What's taking place in this picture?"

Figure 2-6: Originally printed in *Fortune* magazine October 29, 2007 (pg. 130)

The article which accompanied this photo in October 2007's edition of *Fortune* magazine indicated that Nissan technicians were exploring ways to put a door together using as few tools as possible.

The reader might conclude (as can you) that the obvious benefits of their success would be:

- Quicker Assembly

- Less variation

- Fewer needed parts and tools

- Aid in an effort to automate the process

What else can you gather from this picture? As stated earlier, a company that places a corporate and strategic value on equipment reliability is a company that most likely performs many aspects of its business at an exceptional level.

From Nissan's apparent employment of Lean and Six Sigma concepts, it would be a short leap to conclude that Nissan has a level of great competence and has already figured out how to shift from some of the paradigms that we often encounter with poor reliability. Discipline to quality and process is legendary in several Japanese industries. Their culture's acceptance of Dr. W. Edwards Deming's quality teachings and George Smith's reliability learnings bolstered an industrial revolution.

Consider that a task as seemingly routine as that shown in the photo from *Fortune* magazine is really a window into what can be achieved if, and only if, we fully embrace all facets of improvement: Lean, Six Sigma, and reliability. In a sense, we have been in a one hundred year battle with ourselves, and have maintained our position all the while. In contrast, many of our competitors throughout the industrial world have moved on.

Our competitors might know something that we don't.

The Competition

Your competitors, those that are in your industry and are considered the leaders in equipment reliability, like Nissan, recognize that they can't be successful at applying Lean Manufacturing principles if they don't have reliable equipment. Likewise, even Six Sigma efforts without reliable equipment can fail to produce the anticipated results.

Ironically, some people reading this text today are trying in vain to apply Lean Manufacturing and Six Sigma principles without first shoring up equipment reliability. You first have to conquer basic equipment reliability before attempting to improve beyond that. You can't improve on chaos!

Here in the 21st century we are engaged in a global war, a different kind of war, but a war nonetheless. To borrow a military analogy, we can't win a war on the defensive. Good defense might win championships, but it doesn't win a war. What we need is a good offense. The truly fortunate aspect of our path forward is that what's required from us is already known. This is also the most troubling aspect of the paradigm conflict. We know what we are supposed to do, we simply don't do it, or are incapable of doing it.

Conquering both, *can't do* or *won't do* may be as simple as viewing our reliability effort as a whole-team responsibility, and seeing it for what it truly is—a core value business element.

Strategy

Without a doubt, the most troubling aspect of today's maintenance departments (or to a great extent, an overall equipment reliability approach) is that there is no strategy. You're going to prove this right now. In the space below, briefly define your organization's documented (that means it is written down) equipment reliability strategy. Do not write a vision, or a mission statement, but define the officially documented strategy:

_____

_____

_____

_____

Shamefully, most organizations don't have a coherent, well-defined and easily understood reliability strategy; not shame on you, but shame on all of us.

We're going to build maintenance strategy over the next few pages. We'll do this by first laying out the framework of a strategy to follow as a template. Furthermore, we will use this strategy later in this book to demonstrate (as an example) the value of a well-run storeroom, and the importance of having a maintenance planner that actually plans and schedules work.

We are going to use a historical example to develop our strategy template.

Operation Overlord was the World War II code name for D-Day, the Normandy landing of Allied troops on the west coast of France.

This is a very simplified explanation, not meant to be absolutely historically accurate, but to use as a simple guide on which we will overlay our reliability strategy.

The goal of D-Day was to open the Western Front in Europe. The day before D-Day, the Allies had taken Rome, so the U.S. and its allies had a sound foothold on the Southern Front. The Russians were aggressively prosecuting the war on the famed

Eastern Front, and the Scandinavian countries, friendly to the Allies, had found success on the Northern Front. A successfully ensconced Allied force in the West would box the Axis powers in, and contribute strategically to the march on Berlin.

General Eisenhower and his general staff were fully aware that the planning of such an invasion, involving input from fourteen countries, would have to be simple in understanding, and clear in execution. Mistakes would be deadly.

Here is the first element to building an effective strategy. It must be simple to understand, and clear in its execution.

Simple and clear. In WWII, for the invasion of the Normandy coast, assurances had to be made that boats would not collide in the English Channel, nor would planes run into each other in the skies over the channel. Simple in understanding, clear in execution.

Again, this is a very, very basic interpretation of the actual strategy for D-Day, but the approach to developing a strategy is what's most important and is to be emulated.

> *Operation Overlord,* aka *D-Day*
> Goal: Open a Western front in Europe

An Example Of A Strategy

1. Navy frogmen had to clear a path through underwater barriers for the Higgins boats delivering Marines and soldiers to the beach.

2. Navy ships were needed to bombard the fixed German defenses to knock them out or at least soften them up.

3. Allied forces (chiefly the 82nd and 101st Airborne) were to parachute behind the enemy, as well as the accompanying glider forces.

4. Finally, the Allies had to 'take the beach.'

Not minimizing the tremendous loss of life, but imagine for just a moment how the outcome of D-Day might have been different if any of those tactical elements for the overall strategy were missing, or out of order.

If the frogmen, for example, had not successfully cleared the path for the brave soldiers and Marines, what would have happened? Our men would have drowned in twelve feet of water. The German defenses in the surf would have kept the Higgins boats from shore.

If the Navy had not been as on-target and as capable as they were, our armada would have faced everything the Germans had entrenched on the cliffs of Normandy. As it was, the Navy stood offshore, safely out of range, and let go salvo after salvo from their 16-inch guns, and 'marched' their projectiles toward the targets. Fixed defenses become targets. Once the target is found, windage and range are called in to the other ships and all hell breaks loose.

The tactical advantage to parachuting behind the enemy might be obvious. It is the ultimate flanking maneuver. Once behind them, the enemy will have to commit resources to literally shoot backwards. At this point, the Germans defending the coastline had to shoot in front of them and behind them. Also, this tactical move gave the Allies the ability to cut off any retreat, and to guard against reinforcements that could be pouring down from Calais (reinforcements that thankfully never came).

In this brief explanation, we've actually introduced two other attributes of a strategy. Strategies are made up of tactically executable elements, and they have a systematic order.

Now that we have a template to follow, a rather good one, we can use it to develop a maintenance strategy.

This next segment is not intended to be disrespectful to the sacrifices of our forces, it demonstrates how great an example for the concept of 'strategy' that the D-Day invasion truly was. Let's overlay a maintenance strategy alongside Operation Overlord.

Keep in mind what we know so far. A strategy must:

1. Be simple in understanding

2. Be clear in execution

3. Consist of tactically executable elements

4. Have a systematic approach

Overlaying a Maintenance Strategy

This now becomes our maintenance strategy (note that the D-Day strategy is *ghosted out*):

Operation Overlord, aka D-Day
Equipment Reliability Strategy
Goal: Open a Western front in Europe
Goal: Decrease Downtime

Strategy

1. Navy frogmen had to clear a path through underwater barriers for the Higgins boats delivering Marines and soldiers to the beach.

   Execute the most effective and efficient PMs with a target compliance of >98%.

2. Navy ships were needed to bombard the fixed German defenses, to knock them out or at least soften them up.

   Generate corrective maintenance from preventive maintenance, or CM/PM.

3. Allied forces (chiefly the 82nd and 101st Airborne) were to parachute behind the enemy, as well as glider forces.

   Executable work will be planned prior to execution with a target of Planned work vs. Unplanned work >95%.

4. Finally, the Allies had to 'take the beach.'

   Develop and execute a maintenance schedule with a target compliance of >90%.

   The elements in this strategy will be explained in later chapters. What is important to take away from this section on strategy is that you need one. A simple, understandable strategy should make clear what everyone's role is in aiding the overall equipment reliability effort. We all have a role to play in equipment reliability. What's yours?

Record here a four step strategy for reliability improvement. Use the one just laid out as an example. Four steps, simple and clear, in systematic order. Provide, if possible, the level (compliance target) you are hoping to reach for each tactical step (e.g. 98%, 90%).

1. _____

2. _____

3. _____

4. _____

Vision

In Chapter 1, you were asked to document your perceived maintenance vision. The response was to be recorded in Table 1-2 as part of defining your organizational equipment reliability vision.

In this section, we are going to build a vision. We've already sketched out a workable strategy, so now, what is our vision?

Reprint the vision you recorded in Table 1-2 here. We will use this as a comparison. It is possible that you don't currently have a reliability vision. If you do have a reliability vision (this could also be a maintenance vision), find it and see if it matches what you come up with here.

_____

_____

_____

_____

Just a brief note to recognize that for this section, and this section only, we are using maintenance and reliability synonymously. These are not synonymous terms, but what we are drafting is a simple vision for our future of better running equipment.

An actual reliability vision should not be completed unilaterally. It is acceptable, as part of using this workbook, to create a vision alone just to use as an example. The actual maintenance vision for your organization should be created with a small team;

ideally, with those who will benefit from a well-running maintenance organization. Suggested participants might include:

- Maintenance supervisor
- Maintenance technician
- Production supervisor
- Production operator

- Safety
- Purchasing
- Storeroom clerk
- Engineer

Definition

A vision is a glimpse into the future. This vision describes what is
actually happening in the future. Imagine that your maintenance
organization, at this moment, is operating on all cylinders, firing
away, world-class, and everything is working exactly as it should.
The shop looks great, everyone is proactively doing something of
great value, and there are smiles and positive vibes everywhere.

Now, imagine a total stranger walks into your organization
and utters this phrase, "Man, you guys are really _____!"

What is the 'word' you visualize that stranger actually saying?

Creating *Your* Vision

Make a list of ten single words, or hyphenated phrases (like world-
class) that you would love to have used to describe your operation,
practice, or organizational energy at some time in the future:

_____   _____

_____   _____

_____   _____

_____   _____

_____   _____

A vision should be a short, powerful phrase that captures the
attention and imagination of the leader. Long, drawn out vision
statements don't evoke a sense of energy and excitement. For that

reason, we suggest a vision statement that is only ten words long. Ramesh Gulati says, "A vision statement is a short, succinct, and inspiring declaration of what the organization intends to become or to achieve at some point in the future." *(Maintenance and Reliability Best Practices*, p. 29)

Please note, do not use the phrase, "We are going to be... " Forget that! A *vision* is what you are actually 'seeing' in the future. You're not going to *be*, you already *are*.

Use up to three of the ten words you listed to write three draft vision sentences. Each draft vision sentence should be ten words or less in length. Hyphenated words are acceptable. Remember to give the vision some 'punch!'

Record your three draft visions here:

1. _____

2. _____

3. _____

Examples

Here is one vision statement that has always stood out. To this day, this group is still a very successful maintenance organization.

"Empowered associates working in partnership to improve productivity"

My colleague provides excellent insight into parsing the words of a vision statement. He counsels organizations across the spectrum of industries and develops maintenance and reliability strategies to excel performance and add real value. As part of the strategies he helps to develop, my friend guides leaders to understand that words have meaning. His point, and it's a very subtle, yet important one, is that words matter. If a vision is only ten words long, how much space is available to waste on filler? Do people understand what is meant by the simple words listed?

Consider our simple vision statement: "Empowered associates working in partnership to improve productivity"

Words Have Meaning

"Empowered associates." How are they empowered? Who are these associates? What mechanism is in place that gives associates 'power?' Recall the discussion in Chapter 1 defining ownership to indicate responsibility and authority. Empowered associates have well-defined responsibility *and* authority.

Continuing with the idea of "working in partnership," what opportunities exist at your location that allow separate groups to work in partnership? When and how are ideologically polarized groups brought together to work together? Partnership indicates that each side knows what is of importance to the other, and they work for the success of their partner.

In the space provided, list a few examples of where independent groups come together at your facility to work *interdependently* on a goal or objective:

_____

_____

_____

_____

A perfect example of what it means to work in partnership is the image of a team in a three-legged race. In a three-legged race, participants are lashed together at the thigh; one has their right leg tied to the left leg of their 'partner.' The goal is to reach the finish line before the other likewise hobbled teams. You win by working together, starting off on the right foot. Hold it! Both people can't start off with their *right foot*. Think about it.

Each partner knows what the other needs to be successful, and they work to make it happen. Both people get to the finish line at the same time. That's true partnership.

"... improve productivity" has to mean something. To improve productivity, it must first be established what the current rate of productivity is and how it is measured. The vision is to improve productivity, not to be world-class on Monday; maybe a year from Monday, but not this Monday (see Figures 1 and 2 from the introductory chapter). The status quo must be defined and understood by all, and the path forward (improvement) must be clear and desirable.

Speaking your vision should lift the scales from everyone's eyes, so that all can see this new place.

Your Vision

Earlier, you wrote ten visionary words, and from that you drafted three vision statements. From that work, now create a final draft of your vision statement:

_____

_____

Remember, this is just an exercise. Your actual vision statement should be developed by a small team. Use this same process to guide the team effort toward an awesome vision.

As discussed, the words you've decided to use have meaning. Using the example laid out previously, define in unambiguous terms what the words and phrases in your vision mean at your location. Document your response here:

1. _____

2. _____

3. _____

Congratulations, you have successfully drafted a vision statement that has power, energy, and speaks to everyone in the organization, one which clearly owns what the future of your organization's reliability efforts will look like and deliver!

Mission

Recognizing that this might seem redundant, please print your final draft vision statement here:

_____

_____

_____

Every decision, going forward, should be made with this vision in mind. It's not uncommon, during the exercise of dead reckoning, to establish known coordinates and set up the desired direction based on the target or objective. Thus it is with the vision of your maintenance and reliability future. The vision will be your team's focus. The vision will guide you. The vision will deliver you where you intended to be.

Definition

A mission statement is comparable to a vision statement in that it provides understanding and a focal point for others to identify what their role in the effort really is. Recall the definition of a strategy in the opening section of this chapter. The strategy must be simple and understandable. Once they understand the strategy, everyone should identify what their role is in the overall effort.

In our WWII Normandy invasion example, every soldier, sailor, and Marine engaged in Operation Overlord knew the criticality of their executable role and they did it to the best of their ability.

The vision and the mission put into context and meaning exactly what it is 'we' are all about.

The mission statement answers the basic question, "What are we trying to do here?"

The mission statement does this by addressing four specific questions:

1. Who are we?

2. What do we do?

3. How do we do it?

4. Who do we do it for?

A mission statement is much longer than a vision statement, and is not restricted to print that can fit on a bumper sticker or fortune cookie. It doesn't necessarily evoke the same gut-punch a vision statement does, but it serves the vital purpose of tying everything together.

A mission statement can be four sentences or several paragraphs. It must, however, answer the four questions just listed.

Example

Consider using this example as a template:

"We are ACME reliability professionals, working in concert as problem solving experts. We determine the best tools to increase availability and the best approach to guarantee success. We improve today, for the benefit of all stakeholders tomorrow."

See what you can do with a few sample mission statements.

Creating Your Mission

Record sample mission statements below:

Who are we? _____
What we do? _____
How we do it? _____
Who we do it for? _____

Who are we? _____
What we do? _____
How we do it? _____
Who we do it for? _____

Who are we? _____
What we do? _____
How we do it? _____
Who we do it for? _____

Your **Mission**

Choose from what you've written as draft statements to create a workable final copy, and record your mission statement here:

_____
_____
_____
_____

From what you've written as the final draft of a mission statement, answer the following:

Who are we? _____

What we do? _____

How we do it? _____

Who we do it for? _____

Congratulations! You have successfully drafted a workable mission statement for your reliability effort. Of course, as with the vision statement, the actual mission statement should be crafted with your team. The team make-up should be the same characters that were used for the vision development.

Goals

The traditional steps taken before now to develop goals may have been a deleterious effort. The method may have actually hampered any effort to meet the goal. Hasn't it been tradition to establish SMART goals?

Meaning:

- **S**pecific

- **M**easureable

- **A**ttainable

- **R**elevant

- **T**imely

SMART. This simple approach is as dangerous as it is worn out. Like 'synergy' and 'low-hanging fruit,' we need a revolution in our business language. SMART is no longer as smart as once thought.

Our goals, as it turns out, haven't always been that *smart*.

It may even be that our less-than-SMART goals have been forced upon us. Maybe goals should be bottom-up vs. top-down. Here's an example of that epiphany. "We changed the systems so people could measure how they contributed to profitability. We stopped setting goals *for* them. Instead, we worked *with* them to set goals for themselves to stretch sales *and* profit. They've been attaining their goals since that time. We stopped measuring the wrong things. Now we measure, and get, what we want." (J.A. Belasco and R.C. Stayer, *Flight of the Buffalo*, p. 218)

An Anecdotal Example

Years ago, a maintenance organization was under pressure to forward its department's safety goals for the calendar year to the company leadership. The maintenance manager assembled the supervisors for input. One young supervisor suggested that the unit's annual goal ought to be zero recordable injuries for the year. There was agreement among the others present. One supervisor, however, (this book's author), raised an objection. The objection was simple, "It's February 11th, and we've already had a recordable injury this year." The manager asked, "What's your point?"

SMART provides a false sense that whatever is written will be what we go with. SMART gives no credence or address to the means and consequences of what it is the organization is trying to accomplish.

Often, the goal is written as the desired outcome. This is wrong. Here are some examples of poorly written goals:

- Decrease recordable accidents by 25% this fiscal year

- Increase equipment availability by 10% within two quarters

- Decrease stockouts in the storeroom by 50% by year's end

These goals would be typical for any organization. But are they SMART?

- Specific—yes

- Measureable—yes

- Attainable—yes

- Relevant—yes

- Timely—most likely

66   Chapter Two

Where these goals, and SMART, fall short is by not describing the means to attaining this new state. Further, the goal should not state the intended consequence (e.g. decrease by 25%).

A Better Goal

Better written goals might be:

- Decrease recordable accidents this fiscal year

- Increase equipment availability within two quarters

- Decrease stockouts in the storeroom by year's end

SMART isn't wrong, necessarily, it just isn't complete and it provides a very false sense of accomplishment. Simply listing *goals* isn't the same as accomplishing goals. What's chiefly absent is any mention of the *means* needed to reach the desired consequence. In the examples just provided, the consequences are listed along with the goal. As we've discovered, this is not the best way to state a goal. Look at the following examples. The consequence of each goal is printed in italics:

- Decrease recordable accidents *by 25%* this fiscal year

- Increase equipment availability *by 10%* within two quarters

- Decrease stockouts in the storeroom *by 50%* by year's end

Heck, the goal and the intended consequence are the easy part; the 'means' by which we achieve these results takes the real work.

The SMART approach to generating thoughts on goals and intentions is easy and quick. Things that matter, it seems, are things that take effort. Why did we forget this? More importantly, if we don't define the steps necessary to link goal and consequence, how do we intend to rally people to pitch in and "empower associates" to "work in partnership"—borrowing from the vision statement example provided in the vision section.

Imagine linking Goals to Means to Consequences to move fluidly from ideas to action.

Linking Goals to Means to Consequences

Figure 2-7 can help us to understand the sometimes circular flow of goals to means, and on to the intended consequences. This figure is shown as a continuous loop to further visualize the continuous improvement cycle we find ourselves in. Ask yourself this question, "Once we reach the intended consequences, are we actually done?"

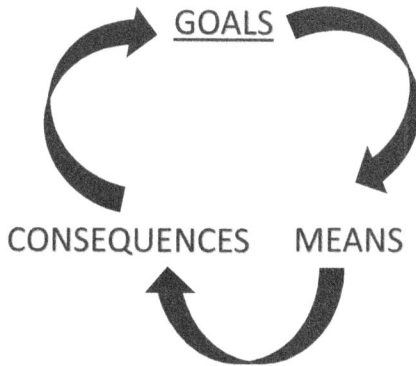

Figure 2-7: Goals to Means to Consequences

Examples of this connection

Goal
■ Deliver Value Added PMs

Means
■ Conduct Preventive Maintenance Workshops

■ Provide Basic Equipment Care Workshops

■ Hire, train, and develop Maintenance Planners and Schedulers

Consequences
■ PM compliance increases to 98%

Corrective Maintenance work generated by Preventive Maintenance activity at a 1:4 ratio

■ Planned work increases to 95%

■ Scheduling compliance increases to 90%

Here is another example:

Goal
  ▪ The Storeroom will have the right part, at the right time, in the right quantity

Means
  ▪ Form, charter, and train a Stores Stock Committee

  ▪ Conduct a Storeroom Effectiveness Assessment

  ▪ Hire, train, and develop Maintenance Planners and Schedulers

  ▪ Provide storeroom training to current storeroom staff and maintenance leadership

Consequences
  ▪ Inventory accuracy increases to 98%

  ▪ Inventory service level increases to 97%

  ▪ Inventory turn rate improves to 1.4

By now you should have the rhythm down, and the flow from goal to means to consequences might come naturally. It helps to imagine one goal at a time.

Give It A Try

Now it's your turn. In the space provided, list three maintenance reliability goals. List only the goals, and not any value (consequence) or process (means) to get to the intended level.

Goals:

_____   _____

_____   _____

_____   _____

_____   _____

_____   _____

Take only one of the goals listed, and spell out the means required to reach that goal. What is it going to take? Don't limit yourself, detail what needs to happen. Here is a hint: there are almost always multiple means to list.

Means required for goal # _____

_____	_____
_____	_____
_____	_____
_____	_____
_____	_____

What would be the intended consequences of achieving the one goal you detailed? Record those consequences in the space provided. Again, there are usually more than one consequence.

_____

_____

_____

_____

Consider using this approach when creating departmental or organizational goals. Simply listing the goal itself, even accompanied by the desired result, doesn't paint the picture of what is needed, nor does it really reach out to your partners to indicate what it will take to reach the goal.

Remember the link—goals to means to consequences.

Sometimes the goal is to complete the fiscal year within budget. That really isn't as easy as it might sound.

Maintenance Budgets

How much did maintenance spend last year? If your company has had a good year, then multiply that value by 10%. That will be your budget for the next year. If your company has had a bad year they are going to need you to decrease next year's spending by 10%.

That might be the easiest, and most common way that maintenance budgets are devised. A simple, absurd example, but it may
not be too far from the truth.

The Genesis of a Maintenance Budget

Where do maintenance budgets come from and how can a real
budget be created and followed? Disclaimer notice: This section
does not cover budgeting for capital or major repair projects.

We should start with a brief mention of the fiscal goal of most
companies. There should be general agreement that for-profit organizations are attempting to make, well, a profit (pun intended). In order to achieve a profit, companies must make a product
or provide a service for less than it takes to make the product or
provide the service.

It would not be unusual for employees in almost any industry to feel that production trumps everything else. Of that there
is little debate. We will put aside the argument that safety is the
number one priority, for the moment, but for-profit industries
really do need to produce to make a profit. This often leads to a
chasm in our dialog.

A Part of the Whole

You wouldn't have to work your way too far up the organizational chart in your company until everyone you encounter is an MBA
or an accountant. These professionals speak a different language,
and it is the language of business: EBITDA (Earnings Before
Interest, Taxes, Depreciation, and Amortization), COGS (Cost Of
Goods Sold), and Conversion Costs (the cost of converting principle supplies into finished goods). Everything that it takes to create
a product or service goes into the cost. Additionally, companies
may add some costs that have nothing to do with the product or
service being created. It would be acceptable for R&D costs to be
shown in the budget as well as philanthropic giving. Budgeting
can be as easy as pie! Figure 2-8 shows the maintenance contribution to costs.

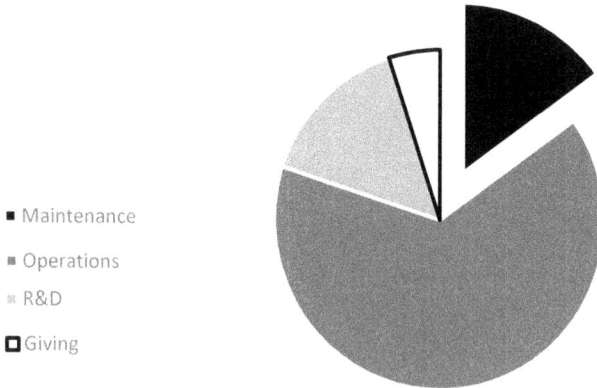

- Maintenance
- Operations
- R&D
- Giving

Figure 2-8: Pie chart showing cut-out costs for maintenance

Figure 2-8 is meant to show the total cost for the company to create a product or service. This could apply to any organization. The cut-out is intended to show, proportionally, the maintenance and reliability costs as compared to the overall cost as an example.

Maintenance Costs

The role of maintenance leadership, quite frankly, is to attempt to walk the budget tightrope through the entire year, and bring in their department with a section of pie no bigger than what was projected.

What is included in maintenance costs? Surely that differs from industry to industry.

Initially, it must be agreed upon as to what exactly makes up the maintenance portion of the costs. Fundamentally, there should be some consensus as to what constitutes maintenance costs. Here are some typical charges that comprise the totality of 'maintenance cost:'

- Maintenance labor
- Overtime
- Overhead
- Contractor costs
- Maintenance training
- Maintenance supplies, including MRO (Maintenance Repair Operations) parts
- Maintenance equipment rental

In the space provided, list the categories of charges that make up the entire spectrum of 'maintenance costs' at your location:

_____   _____

_____   _____

_____   _____

_____   _____

_____   _____

Cost categories, or cost codes, should be easy to understand and communicate.

Budget Categories Explained—and How to Deal With Them

A quick explanation of each of the maintenance cost categories would be helpful to creating and staying within a budget. As with most aspects of equipment reliability, the budgeting process should not be secretive. Of course, personal salaries and compensation packages are kept private, but an important part of surrendering 'authority' and engaging folks in true ownership is to let them know what is going on. A big part of being engaged in the budgeting process is being in on the money discussion.

A perfect analogy to this is remembering back to when most of us were young. As young kids, we had no idea how much money our mothers and fathers made. As you got older, you began to get a sense of how much money the family brought in each month. Pretty soon, you were getting an allowance, and/or possibly you had a part time job. As a teen or young adult, you were more inclined to be 'engaged' in family money discussion (e.g. vacation plans, holiday gift giving, etc.). So it is with maintenance budgets. We can't expect our maintenance associates to help stay within a budget if they are ignorant of the budget process.

Labor

Labor is pretty straightforward. Unless otherwise designed, labor is comprised of forty hours of pay, per hourly person, based on the hourly rate they are due. This is simple, but it is one of the most important elements to keep in mind when trying to stay within budget. It is usually undesirable to cut cost or reduce spending by cutting labor (people). This should be the absolute last step taken to reduce cost.

There is enough trouble finding good skilled labor these days. It would be questionable leadership to cut skilled labor from the work force. You may never get them back.

Overtime
Labor costs usually are fixed during the year; it is the overtime that can be a budget killer. Overtime for hourly personnel or non-exempt personnel could be a 'go-to' category when looking to cut costs. With that being said, there is no greater guarantee for overtime requirements to increase than for management to post a notice that states, "Effective immediately, there will be no overtime." Maintenance technicians are experts at rationalizing the need for overtime.

Overtime, itself, is not bad; in fact, it is necessary for a company that is really growing and has a product or service that is in demand. Overtime is the accumulator of human resource ability that helps the organization get by without having to hire and, later possibly lay-off, employees.

Overtime is one budgeting element that we can and need to control. Ideally, overtime targets should be 10% or less of total labor hours. This would be a very good time to ask how overtime can be controlled. What causes overtime?

To keep this from being a book on just overtime, it should clearly be stated that overtime, or more specifically, maintenance overtime, is often a direct result of something unexpected occurring. Those events might include:

- The customer wants/needs the product now

- Equipment failed during production, leading to off-hour production runs

- Higher demand than anticipated

- Equipment breakdowns cause maintenance shifts to stay over to fix the machine

The key, the absolute *key* to controlling overtime is controlling what creates overtime events. All the causes listed are symptoms and not root causes. There are many truths in our business, and one of them is certainly that the ability of maintenance to stick to a maintenance schedule is never better than production's ability to stick to a production schedule. Maintenance needs to do everything in its power to help production stick to its schedule.

What are some other reasons, perhaps in your company, that might cause maintenance overtime?

List them here:

_____   _____
_____   _____
_____   _____
_____   _____

Are these symptoms or root causes?

Maintenance Supplies, Including Spare Parts

Another budget category is maintenance supplies, including spare parts. It's not unusual for up to 30% of a company's expenses to be equipment reliability related, and also not unusual for up to 40% of that cost to be spare parts related. We are talking about a very significant amount of money. Therefore, this particular category, centering on spare parts, is usually a high-valued target for budgeting and cost cutting efforts.

It's true that you won't cost cut your way to world class. It's also true that we often spend money on spare parts that never get used for the purpose they were intended. A primary contributor to the amount of spare parts stocked and used in most industrial settings is our overwhelmingly poor troubleshooting.

There is a straight line connection between poor troubleshooting and parts usage. Poor troubleshooting can be caused chiefly by inexperienced technicians, trained but poorly intentioned technicians, inadequate troubleshooting time, and improper tools/techniques/guides to troubleshooting. Can you list any other contributing factors to poor troubleshooting?

_____   _____
_____   _____
_____   _____

The spare parts and materials category in the maintenance budget has to be controlled. Unfortunately, there is no easy road for this. Troubleshooting and parts usage begin at the beginning. Equipment must be designed with reliability and maintainability in mind, while still in the concept phase.

OEMs (Original Equipment Manufacturers) will give you the part they get the best 'deal' on if you don't provide them with a list of the exact parts you want on your new machine. How robust is their part going to be in your aggressive environment?

Maintenance technician training, storeroom control, and maintenance planning and scheduling are the other major elements required to get on top of spare parts and materials usage. Each of these topics are discussed later in this workbook.

Overhead

A likely target for budget cuts, and one that generally always opens a less than *genteel* discussion, centers around overhead. For this discussion, overhead is comprised of those people who are exempt from receiving overtime payments. This classification of associates describes a person that is needed for the business, but there is no direct line linking their presence to the production of a product or service. Their work isn't directly connected to profit. That may be harsh but it is very likely to be true. And more than likely, it is the reason this category of maintenance cost is often targeted. Better yet, often times these roles go unfilled because of cost. The savings come from attrition.

Unfortunately, some of the roles mentioned in the previous paragraphs are also relevant here. To reduce the cost of maintenance, and to get more work done, an organization needs a planner/scheduler, needs people in the storeroom, and requires effective supervision. Most of these positions are traditionally overhead.

Contractors

Very few organizations can operate without maintenance contractors. In fact, this author has only ever seen one such example that even came close. The result was a swollen maintenance organization that had technicians literally sitting around until their specific skill was required. Contractor costs are a necessary hit to the bottom line, and the cost has to be controlled.

In fact, due to some understandable regulatory and safety rules and policies, the use of contractors is often spelled out. Overhead crane maintenance, for example, would be an area where a company would be well within their safety-centric best interest to contract the work to those who are experts in that field.

Regardless of location, most companies and maintenance groups have a go-to contractor in their area, or for specific purposes (e.g. ammonia compressor rebuilds) that they use all the time. These very valuable augmented forces need to be part of the equation when it comes to establishing this critical budget category. For this particular element, you do tend to get what you pay for. Who is your go-to contractor?

_____

_____

_____

Contractor costs are a very important part of a maintenance budget. These costs are often cited as being a significant reason that the maintenance budget was exceeded. It is very difficult to stay within a budget set for contracting costs because we budget for planned contractor work, when it is the unplanned work that breaks the budget. Last minute costs due to mobilizing an army of contractors are hard to recover from.

Maintenance Training
Without a doubt, as a failure across the institution of maintenance and reliability, we've all done a poor job at securing maintenance training resources (money + time) and establishing a skills matrix and gap analysis. Certainly, some of you who are reading this have done a great job, but as a whole, across the spectrum of all of us, this is a lacking element. Almost every aspect of the maintenance budget, from establishing it to staying within it, is affected by how well (therefore, how capable) our team really is. This is a category that we most likely have to fight to save.

Figures as high as one hundred hours per skilled-trades person are required for effective skills training. Maintenance training must be skill based and not just general training. Typical general training is often necessary, but can lead to a false conclusion that 'our maintenance folks get a lot of training.' Such general training might include:

- HAZMAT
- HAZCOMM
- Fork Truck

- Safety
- Fire extinguisher
- First aid

- CPR

These are important and necessary subjects, but do not count towards one hundred hours of skill based training.

Since our goal is to provide this skills training, we simply multiply the number of technicians we have by one hundred hours, times an average of their pay, to get a total of dollars in hourly pay attributed to training costs, and add the additional cost of the training.

Training can be accomplished through vendors on-site, off-site training, college or technical schools, or sending technicians away for specific training. For example, Variable Frequency Drive and Programmable Logic Control (PLC) training would be common off-site, distance learning opportunities.

The key to establishing a training budget is to establish the training need first.

Maintenance Equipment Rentals or Leases

The final category for maintenance budgets to address is the maintenance equipment rentals or leases. This is generally a pay and forget category. Not until we really start to scrutinize the maintenance budget does it become clear exactly what we own and what we don't. Repair and replacement costs for rental and leased equipment are often a contributor to budget overruns.

Recalling the thousands of dollars spent on fork truck repairs and man-lift maintenance can certainly give the reviewer a sense of money not well spent.

How does establishing a well thought out and objective budget compel company leaders to understand the vision?

Taking Control of the Maintenance Budget

An effectively managed maintenance organization will find a way to reduce maintenance cost 3% per year. Now really think about that. Three percent per year. There are a few absolute truths about maintenance budgets, the most important being, "If we don't control maintenance costs, someone else will." It's very likely that other person will not have the best interest of the maintenance department in mind.

At the beginning of this section, we listed several maintenance budget categories as an example:

- Maintenance labor
- Overtime
- Maintenance supplies, including MRO parts
- Overhead

- Contractor costs
- Maintenance training
- Maintenance equipment rental

We only want to consider a few of these categories as we look for ways to cut 3% per year.

For example, the following categories are those that should be safe from a reduction in funding:

- Maintenance labor
- Overhead
- Maintenance training

That leaves the remaining categories as those that can and should be looked at for an annual costs savings.

We're looking at maintenance budgeting as a block category, a piece of the pie that we spoke about earlier. There are, of course, several methods used to determine how to divide up the monies. But first, it is necessary to determine how much money we should set aside for maintenance.

Let's use your numbers as an example. Earlier, you were asked to list the categories of charges that make up the entire spectrum of 'maintenance costs' at your location.

An Exercise in Maintenance Budgeting

Table 2-2 provides space to list those categories again, as well as the current known budget value for each category. Just fill in the Current Year Budget Value at this time.

Table 2-2: Maintenance budget

Category	Current Year Budget Value	Next Year's Budget Value
TOTAL		

You may have listed categories such as labor, overhead, and training. These categories are generally the specific ones that we want to guard as much as possible. A cut in funding for these particular categories would be detrimental to a reliability effort and morale. For budgeting purposes, as stated, these are categories we want to protect. The remaining categories are the ones we can manipulate to reach a desired reduction of 3%.

Use the example shown in Table 2-3 to complete the budget you started in Table 2-2. Protect the categories we spoke about,

Table 2-3: Sample maintenance budget

Category	Current Year Budget Value	Next Year's Budget Value
Maintenance labor	$8 000	$825,000
Overhead	$ 500	$175,500
Maintenance training	9 000	$95,000
Parts and material	$ 056,0	$1,984,080
Overtime	$ 500	$16,100
Contractor costs	8 000	$200,000
Eqi pment rental and lease	$ 000	$45,000
TOTAL	$ 444,0	$3,340,680

as much as you can, and show that you've created a budget that results in a 3% reduction in expenditures. Go back at this time and complete Table 2-2.

I promised this would be a very simple example. Let's rationalize some of the entries.

Consider that in this example, there are three categories we want to protect. As mentioned before, those are labor, overhead, and training. It's very likely that these won't actually hold steady, as there is a necessity to account for wage increases, new hires, and bonuses; these are likely to increase in practical application.

The value shown in the total row of $3,340,680 for next year is a 3% reduction of the overall budget from the current year. Note too that next year's value for rentals has remained the same; this would not be unusual as this is commonly a contracted service.

The exercise now becomes a drill in arriving at values for parts and materials, overtime, and contractor costs that will result in the desired goal value of a 3% reduction in overall costs.

If you have not done so, please complete Table 2-2 with your categories and your numbers for the current year and the next year.

At the start of this section, I facetiously wrote, "How much did maintenance spend last year? If your company has had a good year, then multiply that value by 10%. That's your budget for this next year. If your company has had a bad year, they're going to need you to decrease next year's spending by 10%." This isn't really how maintenance budgets are made but sometimes it feels like that is exactly how it is done.

What We're Really After With The Maintenance Budget

Our goal should always be to drive the maintenance portion of the pie (see Figure 2-8) to be a smaller percentage of the overall pie. Similar to most aspects of high performing operations, we want to drive this reduction in a slow, methodical, and strategic manner. The goal of 3% was offered earlier, and so it is recommended that our offering each year should be a reduction of 3%; keep in mind that some categories of our budget may, in fact, increase.

Now that we've arrived at a workable number for next year's budget, let's understand the different ways to execute an annual budget. There are two core methods: history based and zero based.

History based budgeting is most likely what most readers are familiar with. We've actually been referring to this process throughout this section. In history based budgeting, the previous year's budget is used as a guide to the development of the new budget. Increases are made as costs are most assuredly increased over the course of a year. This is generally an acceptable process. It is certainly the quickest.

The only negatives to this traditional format are the lack of consideration for the actual spending, and the fact that it goes against the world-class idea of decreasing maintenance costs per year.

Companies have compensated for part of shortfalls by addressing actual spending in some maintenance categories, but leaving off the responsible 'need' to reduce overall costs. The major downside to historical budget setting is making it through an entire year and having any money left. It could be all gone in the first few months.

Zero based budgeting begins by allocating a pot of money for each period. Most companies using this approach divide that money into periods of the year, of which there can be up to thirteen, or by months. If this is the approach, it is important to advise those holding the purse strings that the spending could be different for every period. It is unlikely that all expenditures for each period or month would be exactly the same. The downside to zero based budgeting is primarily that you have to use it or lose it. If the money is not spent for the clearly approved and justified reason, it is lost at the end of the period.

These two, primary core methods of budgeting have led to great creativeness in operating and executing maintenance budgets. Company or plant leadership will often ask, "Can we move that major repair to next quarter (month, period, week, etc.)?" The purpose is simply, "We're out of money this quarter (month, period, week, etc.)."

Some companies have even gone so far as to capitalize major repairs (which would normally be expensed), justifying this deed as 'life extending' work. I'm not an accountant, and people a lot smarter than me on this sort of P & L/Balance Sheet rationalization have clearly said this was acceptable and legal.

We've even gone so far as to defer maintenance to funnel monies toward more critical, production-related needs. The example of repairing a plant roof is always a good analogy to use to prove this point.

Budgeting is tough. It's hard to arrive at the correct number as to what the bar should be set at for next year and the following years. It's even tougher to make it through an entire year and stay within budget. Not impossible, just tough.

I want to say a few words about why this is so tough. Perhaps the most significant reason, and this is hard to hear, is that maintenance and reliability in general are not seen today in most facilities as an asset to the business. We've worn out the phrase that "maintenance is seen as a necessary evil." Here is the hard, straight truth. Most maintenance organizations are treated as a cost to the bottom line. They are viewed as an expense. Maintenance costs the organization money, but does not have anything to show for the investment.

Consider that on the production side, there is a budgeting process as well. There is a cost to buying, transporting, and warehousing the principal supplies. Production has labor and overhead costs, not to mention overtime. Production training is not to be left out, nor should the other costs that production records in their piece of the pie.

There are many similarities, but one glaring difference. Production takes those costs, if you will, and creates income from them. They provide a product more valuable than the sum of the costs that go into the product itself, with the product being an actual product or perhaps a service.

The Struggle Is Real

Maintenance has the same costs. The *product* maintenance provides is the *service* it provides. Today, however, that service is not seen by others as being of more value than the costs that it takes to 'make' the service. That results in the constant pressure to reduce the costs of maintenance, or just as bad, an arbitrary cut in spare parts.

In order to stick to a maintenance budget, the following value added activities are required:

- The right part, at the right time, in the right quantity

- Access to the capital equipment for preventive and corrective maintenance according to the maintenance schedule

- Competency based, and skill based training (one hundred hours per person, per year)

- Highly effective and efficient preventive and predictive maintenance

- Maintenance planning and scheduling

- A high performing maintenance storeroom

As stated earlier, you are not going to cost cut your way to world-class. However, if we work together, purposefully and strategically, we can reduce the maintenance piece of the pie every year, while providing more and better equipment reliability.

The Really Important KPIs

In the space below, list all the maintenance and reliability related metrics and Key Performance Indicators (KPIs) currently tracked at your location:

_____   _____

_____   _____

_____   _____

_____   _____

_____   _____

We often use metric and KPI synonymously. The trouble is, they often don't mean the same thing. In fact, the terms themselves should provide direction on how much we care, and what we should do based on the results of the measure.

How are metrics and KPIs different?

Definitions

A metric is often a measure of something we take at a set sampling period to determine its magnitude. This is typically a number (or value) that we track to gain a sense of the trend of some activity.

A KPI, or Key Performance Indicator, is a type of metric that is tied directly to a stated (and therefore, written down) goal or objective.

Figure 2-9 graphically shows the relationship between these two sets of measures.

Following is a useful analogy to clarify the difference between a metric and a KPI.

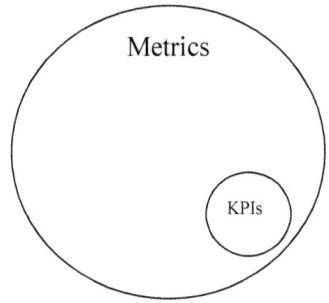

Figure 2-9: The relationship between metrics and KPIs

A Practical Example

A person may want to lose weight because they're feeling sluggish or their clothes are a little tight. They would start by getting on the bathroom scale and making note of their initial weight. They would spend days and weeks, perhaps months, eating less and exercising more. Weight loss, after all, is simply a process of less calories in, more calories out. Each week they would get on the scale and track their progress. They would be advised by friends to weigh themselves no more than once weekly, and weigh themselves on the same day, in the same manner of dress as they had initially. That will give them a true apples-to-apples comparison when tracking their progress.

In Figure 2-10, the weight loss effort is tracked using the metric of pounds lost per week.

Metrics differ from KPIs specifically in this way: Key Performance Indicators are metrics which are directly tied to a stated goal or objective. Here is an example where tracking weight loss is no mere metric, but is now tied to the stated goal to lose weight, and is now a KEY Performance Indicator.

If a person needs to lose thirty-five pounds in seven months because their high school reunion is coming up, they would decide

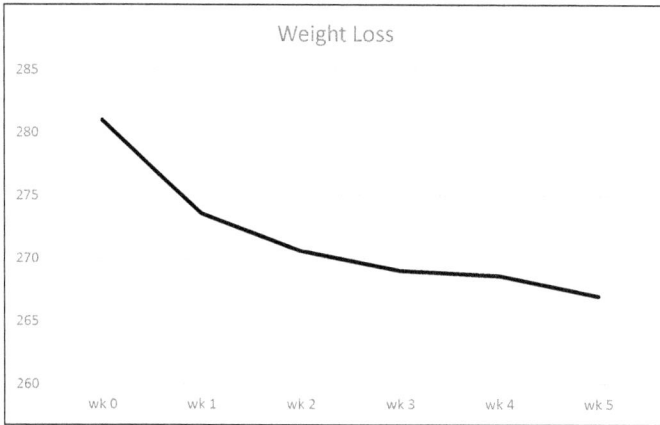

Figure 2-10: Weight loss chart as a simple metric

that they need to lose, on average, five pounds per month. They would develop a plan to eat less and exercise more. They would sketch out a strategic plan that included what to eat, drink, and how to fit exercise into their busy schedule. They would also, as in our example above, get an initial weight, and then vow to measure every week at the same time.

What is different about these two scenarios? In the second example, the person seeking weight loss has made a declaration to lose weight for a specific cause—a stated goal or objective. When our subject gets on the scale after the first month, and has indeed lost five pounds, they will continue with their plan. If they get on the scale the second month, and have only lost four pounds, they will set a course to lose six pounds the third month, and so on.

A KPI, therefore, causes us to alter our course in an effort to put us back on the path if we were to stray. It often points to what went wrong, or what we did wrong to veer off course. It will also tell us if our goal or objective is attainable, and if our strategy was lacking to begin with.

Figure 2-11 on the following page, shows weight loss as a KPI.

Is it helpful to know the difference between the uses of a metric and KPI?

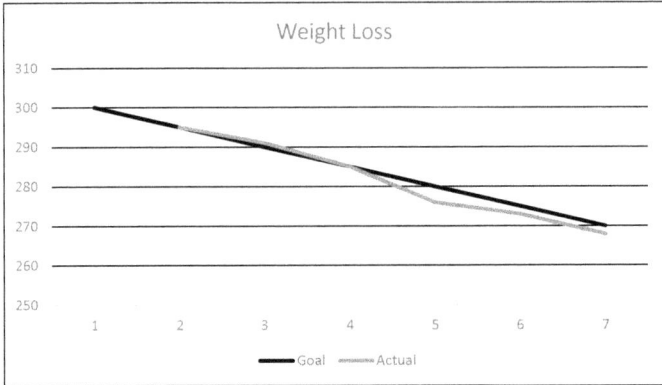

Figure 2-11: Weight loss as a KPI

Knowing The Intended Purpose

It was important to initially provide some distance between the meaning of a metric and a KPI, and to ensure that we refer to these two measures with the intended purpose. A KPI is tied to a stated goal or objective. For professional purposes, those goals or objectives are always written down. Also, the action to take if the trend is heading in the wrong direction is communicated and known to everyone. Determining an activity, assigning a goal result, and providing guidance to stay the course makes an activity a true strategic initiative.

Regardless of tracking metrics or KPIs, at some point it becomes necessary to determine which of the almost 155 maintenance and reliability measures are most effective, and which of these best tell the story of the maintenance efforts and history at any location.

The following is a non-exhaustive list of measures normally seen in production environments:

- Downtime
- Uptime
- PM Compliance
- Scheduling Compliance
- Emergency Work Orders
- Mean Time Between Failure (MTBF)
- Overall Equipment Effectiveness (OEE)
- Stockouts
- Inventory Accuracy

- Recordable Injuries

- Pounds produced

- Scrap

Just choosing and posting a graph is not enough.

The Cardinal Rules For Posting Information

There are a few empirical rules about selecting and posting metrics. They are simple rules, but mightily important.

- If possible, select a measure that measures the 'thing' you want

- All posted metrics will be displayed as a chart or graph

- If you post a metric, be able to explain to everyone what and how the metric is calculated

- If you post a metric, be able to explain to the actual people you're 'speaking to' how to improve or change the value

We are going to review these elements in greater detail.

Measure The 'Thing' You Want

There have surely been countless studies on the power of suggestion. Even management training courses recommend the virtues of optimistic outlooks. Dr. Norman Vincent Peale wrote a book on the "Power of Positive Thinking," and so it should be with our metrics and KPIs.

Consider the unique case of uptime vs. downtime. They both give us an idea of the same value, which is 'how available is my equipment for production?' When given an option, consider what the graph itself is going to look like, and what message you are hoping to convey. We want an up arrow to project the positive, meaning that a down arrow type graph in the mix can be confusing. Sometimes down is good, and the graph should clearly point out this fact.

This sounds elementary, but give some thought to the fact that you want to always encourage your team to understand that 'up' is good, and we need to do more of the things that make this arrow go up. As just mentioned, if forced to, use a graph where 'down' is good and clearly make this point unambiguous.

Figure 2-12 is a simple graph that could indicate the trend in uptime or downtime.

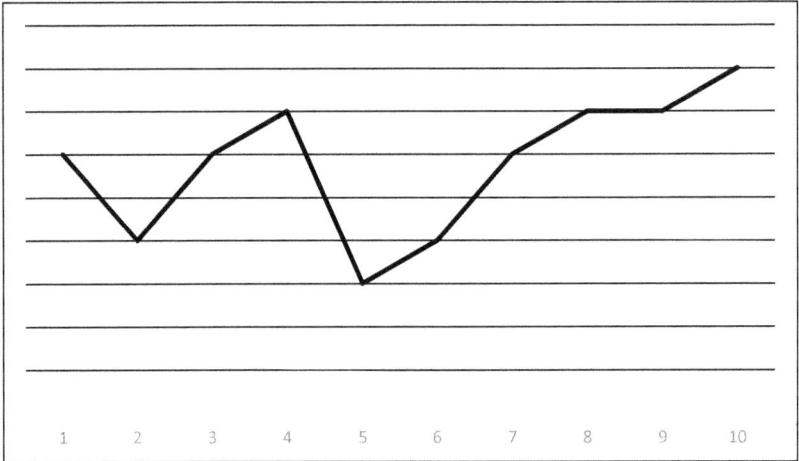

Figure 2-12: Does this graph show uptime or downtime?

Without any other markers, the viewer would be hard pressed to determine if the current course of action should be maintained, or if something significant needs to improve in the maintenance department. Figure 2-12 does little to communicate if "we're doing a good job," or not.

Figure 2-13 represents the inverse of the information in Figure 2-12. The same shortcomings are true in this graph as well.

Of course this is a trick question, but which of the two graphs (Figures 2-12 or 2-13) displays uptime, and which displays downtime? These two graphs are reporting the same result, each week; one is uptime and one is downtime. Which is which?

The actual display of information can determine if anyone is even going to look at it, much less be able to interpret it.

A lot of the graphs and postings on plant and facility bulletin boards have become little more than white noise. We need to

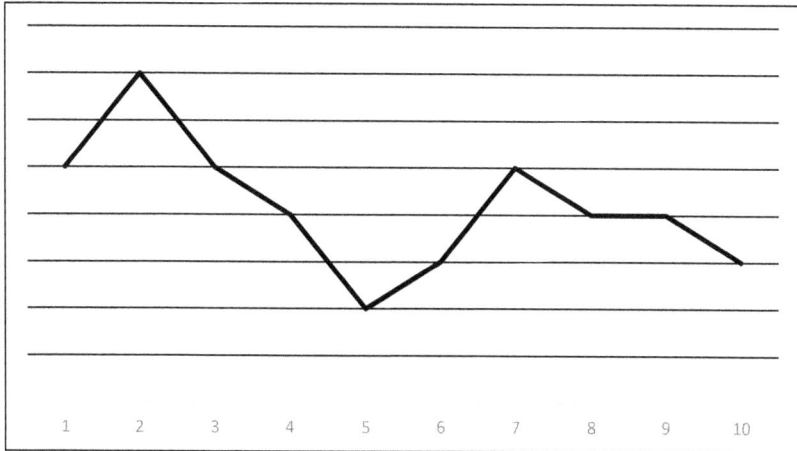

Figure 2-13: Does this graph show uptime or downtime?

leverage a nuance of the human brain, which is that we all think in *pictures*.

Post All Metrics as a Graph

This 'thinking in pictures' is why we display information in the form of a graph. Very few people can convert a table to a graph in their mind, or on the fly. We must adopt a philosophy that, whenever possible, visual data communication (charts, graphs, etc.) will be displayed in such a way to convey the 'thing' we want more of. Also, as a graph, the image will quickly convey a message to the viewer that the metric is heading in either the right or the wrong direction. When possible, 'up' should be the desired direction. Like that old Johnny Mercer song, "Accentuate the positive."

Twelve traditional metrics were listed a few paragraphs earlier. Here is the same listing accompanied with a more positive suggested measure, where applicable.

1. Downtime—UPTIME

2. Uptime—UPTIME

3. PM Compliance—PM COMPLIANCE

4. Scheduling Compliance—SCHEDULING COMPLIANCE

5. MTBF—MTBF

6. Emergency Work Orders—PLANNED VS. UNPLANNED

7. OEE—OEE

8. Stockouts–INVENTORY SERVICE LEVEL

9. Inventory Accuracy—INVENTORY ACCURACY

10. Recordable Injuries—INJURY -FREE WORK HOURS

11. Pounds Produced—PRODUCTION SCHEDULE COMPLIANCE

12. Scrap—% QUALITY PRODUCED

At the start of this section, you were asked to list the equipment reliability and maintenance related metrics and KPIs currently tracked at your location. Table 2-4 provides an area for you to list those measures one more time, in the relevant column. Where applicable, list the complementary and more positive measure. Finally, indicate if the measure currently tracked is a metric or a KPI. Remember, a KPI is directly connected to a stated (written down and communicated) goal or objective.

Table 2-4: Current vs. Complementary measure comparison

Current Measure	Complimentary Measure	Metric or KPI

It was mentioned earlier that "people think in terms of images." This is true. I am going to write a word below, and I'd like you to write the image that pops into your mind. Ready?

"MUSTANG"

What image popped into your mind?

_____

_____

_____

_____

This is an example that I've personally used around the world. Typical responses include:

- Car

- Ford, the company logo

- The mustang image on the car

- 5.0, the block lettering on the side of a car

- P-51 (the WWII airplane)

- Red

- Convertible

- Some people even hear a tune, the old Mack Rice song, *Mustang Sally*

Interestingly, this exercise performed for a class in Trinidad resulted in everyone shouting "candy bar." I was informed that a Three Musketeers bar (Mars, Inc.) is called a mustang on this South Caribbean isle.

If we think in terms of images, it would stand to reason that images must have meaning.

Be Able To Explain The Meaning And The Calculation To Everyone

We all think in terms of images, not tables, or lines of text. It should follow, therefore, that when we communicate maintenance and reliability measures we should do so in the form of a graph. The eyes are drawn to the beginning and end points; the trend line gives an instant idea of an established track record over time. Not that results are this instantaneous, but imagine a person performing a task and 'seeing' the line moving up the graph. That is the connection we want.

This is going to sound almost unbelievable, but it is remarkably true. For all the measures currently taken, charted, trended, and acted upon, there is a very good chance that no one can actually explain what the measure is measuring, or how it is calculated. It is more likely that there are two or three, loosely-related definitions at play.

Don't believe me? Try this. Ask three people in your organization what 'downtime' means. Don't lead their answers, and ask them where the values, which make up the measure, come from (the numerator and the denominator). Record those answers here:

_____   _____

_____   _____

_____   _____

Just for fun, compare their answers to your answer for a similar question in the introductory chapter of this workbook. It is highly probable that there are three or four differing interpretations of this very common measure. Isn't it interesting, and a little tragic, that so much is invested in a number (like downtime), yet we don't really know what it means, where it comes from, or how to change it?

What can we do?

For every chart or graph that you are responsible for creating and posting, it is your responsibility to explain (or be capable of explaining) to everyone who might see that post what it means.

Figure 2-14: Car dashboard

Here's a common example of this. In Figure 2-14, can you explain what the highlighted gauge tells us?

The highlighted gauge in Figure 2-14 is the speedometer on a car's dashboard. Almost everyone is familiar with this gauge. If you had to, you could inform those who see this device as to what it is telling you. For example, you might tell them that it tells you how fast you are going. There is no need to get over complicated on the definition; remember, this is an education for the masses, meaning all those who might see this measure.

Further, if necessary, you could provide some general information as to how the value is generated. What are the inputs and outputs? Again, be as specific as necessary to convey the point. The real detail will come with and to those who are using it as part of their role and function and whose work is being measured.

Now, imagine a scenario where it was explained to you, as a casual observer, what a speedometer tells you, and generally how that measure is derived. Imagine now that you are asked to drive the car. Without any familiarity beyond basic driving skills, if you were asked to move the speedometer needle up or down, what would those instructions consist of?

Explain How To 'Move The Needle'
This is an explanation of the point made earlier about measures. Namely, that we need to explain, or be able to explain, what is necessary to move, improve, or in this case, change the value. This is necessary for the actual people we are 'speaking to' with this graph.

If you were in the driver's seat, and I was in the passenger's seat, and we were on the road, there are a few actions I could advise you to take to move the speedometer needle 'up' and make the car go faster:

- Press the accelerator

- Go down a hill

Likewise, there are few actions I could tell you to perform in order to move the needle 'down,' make the car go slower:
- Take your foot off the accelerator

- Take your foot off the accelerator and press the brake

- Stop completely

- Stop and turn the car off

- Go up a hill

The plain truth of the matter is, when it comes to maintenance and reliability measures, we always need to explain what
the measure is measuring, how it is measured (where the values
come from), and what must be done in order to move the needle.
If we do not, or cannot perform these basic leadership tenets, we
are abdicating the definitions to popular opinion, and the actions
to take are left up to people to figure out for themselves. Their
actions might not be the actions we want them to take.

Here is a practical example: we want to measure our ability to
plan, schedule, and execute a preventive maintenance program.
Essentially, we would like to measure how effective we are at
getting the PM work orders completed on time. The graph we post
might look like Figure 2-15.

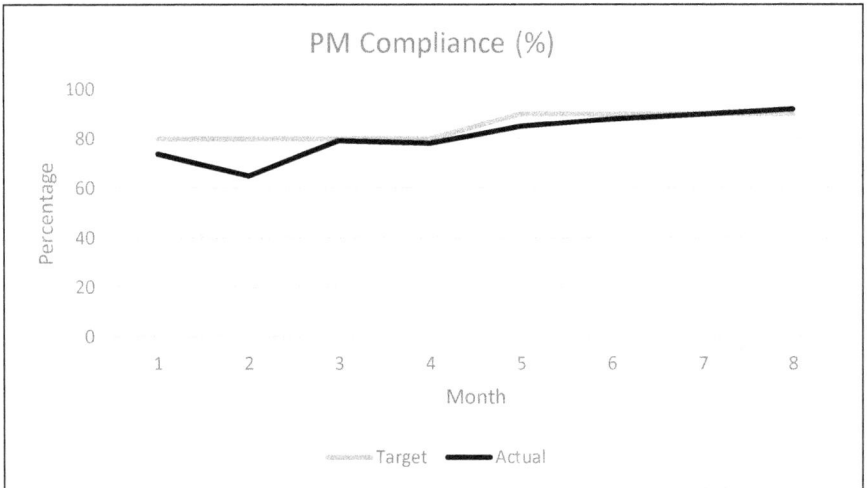

Figure 2-15: Sample PM Compliance graph

Here's our pitch:

Folks, this graph represents a measure of the PMs we completed vs. those PMs that we scheduled. PM compliance is measured,
by the planner, each week, and then the weekly compliance values

are summed up for a monthly total. Please note, the PMs that were accomplished must be the same PMs that were scheduled. This is a sum total of weekly, monthly, quarterly, semi-annual, and annual PMs that came due during this timeframe.

The formula for calculating PM Compliance is:

$$\frac{\text{PMs Completed} \times 100}{\text{PMs Scheduled}}$$

PM Compliance is shown as a percentage.

Anyone who wonders what the PM Compliance graph is all about should be greatly educated on the subject after this explanation.

However, we don't expect everyone to be engaged in the PM program, not directly. Even less may actually care. But there is a specific audience we are targeting; namely, the maintenance technicians, our production partners, the storeroom, and maintenance planners/schedulers.

If you were consulting with your maintenance team, and the question "What can we do to improve (increase) PM compliance?" was asked, what would your advice be? Take a shot and record your advice in the space provided.

Team, here are some actions I'd suggest we take to improve our PM Compliance:

_____

_____

_____

_____

Here are some suggestions we came up with:

- Scrub the PMs that we do to minimize the CYA (Cover Your Ass) PMs, and eliminate the non-value PMs

- If you, the technician, feel a PM task is not necessary, or done too often, please let me know

- When building the maintenance schedule, first add the PMs which are due

- If you get a PM assignment, be on time at the equipment and have all the tools and supplies you need for the job

- If you get to a machine, and the operator or production supervisor won't give you access to complete the PM, call me (the supervisor) immediately

- If you can't get to a machine because you're tied up on another job when it's time to perform the PM, call me immediately

- If someone wants you to wrap up your PM quicker than planned, let me know immediately

- If you discover something during the PM that needs immediate attention, call me immediately

- If anything keeps you from being at the machine, equipped and ready to go, or keeps you from performing the PM task as planned and scheduled, let me know immediately

This section is titled RELEVANT KPIs, so this would be a good time to discuss what KPIs would be relevant.

Positive and Relevant KPIs

Key Performance Indicators are metrics which are tied directly to a stated (and therefore, written down) goal or objective. Earlier in this chapter we worked at putting together a strategy:

Equipment Reliability Strategy
Goal: Decrease Downtime
Strategy

- Execute the most effective and efficient PMs with a target compliance of >98%

- Generate corrective maintenance from preventive maintenance, or CM/PM

- Executable work will be planned prior to execution with a target of Planned work vs. Unplanned work >95%

- Develop and execute a maintenance schedule with a target compliance of >90%

After our discussion on linking goals to means to consequences, we can better define our equipment reliability strategy as:

- Create and execute value added preventive and predictive maintenance

- Generate corrective maintenance jobs from preventive maintenance work

- Plan maintenance work

- Schedule, to the hour, planned maintenance work

This strategy, using the goals to means to consequences approach will lead to the creation of the goals and objectives. Whereby, quite literally, this becomes the 'stated goals or objectives.'
From earlier in this chapter:

Goal
- Deliver value added PMs

Means
- Conduct preventive maintenance workshops

- Provide Basic Equipment Care (BEC) workshops

- Hire, train, and develop maintenance Planners and Schedulers

Consequences
- PM compliance increases to 98%

Corrective Maintenance (CM) work generated by preventive maintenance activity at a 1:4 ratio

- Planned work increases to 95%

- Scheduling compliance increases to >90%

This is the stated goal. As such, our measures for these elements will be considered Key Performance Indicators.

Our KPIs, for this specific goal are:

- PM compliance
- CM from PM
- Planned vs. Unplanned
- Schedule Compliance

It's no accident that these metrics look familiar. They are not only the chief measures used to demonstrate the results of the maintenance effort, they are truly some of what I call the Sergeant-Major metrics. These are the top echelon of values.

Of the nearly 155 maintenance metrics that exist across the spectrum of industries, we only need to select a dozen or less to really tell the story, and clearly (and concisely) relay the results of the maintenance and reliability sweat equity.

Aside from the four KPIs listed above, there are a few other KPIs and metrics I'd recommend to round out the offering of positive and relevant KPIs:

- MTBF (Mean Time Between Failure)

- MTTR (Mean Time To Repair)

- Inventory Accuracy

- Inventory Turn Rate

- Inventory Service Level

Take these closing comments with a grain of salt. No doubt you've noticed that I did not list two of the more common maintenance productivity measures. I usually caution against posting downtime or uptime for the simple reason that there is never an absolutely clear definition of what these measures mean. Nor does it seem that maintenance has any control over these values. It's very likely that production owns the *button* used to enter equipment availability. Said another way, a maintenance person cannot usually look at a downtime chart, note a trend in the wrong direction, and know intuitively what to do to move the needle.

I'd warn against having more than a dozen measures. Ever! As much as possible, I'd recommend tying your measures to a goal, thus making it a KPI. Always include clear actions to take when the trend is going in the wrong direction.

Change Management

People don't hate or dislike change so much as they don't like to be surprised by it. In fact, very little has not changed in our lives. It's likely, if you're an adult male, the only thing in your life that hasn't changed is your name.

What is generally resisted is change that catches us off guard. We need time to process and to determine how this new order will affect us. There is absolutely nothing wrong with taking a selfish moment to determine how to respond to change and how to operate with the new set of rules.

Earlier, we discussed a common leadership desire to change the culture in an organization. Culture can't really be changed. In fact, it may not be what is really needed. However, if it is needed, the philosophy has be to slow and pragmatic, and must follow a transparent process. Gulati reckons, "No matter how many goals are set or how grand the vision, an organization can go only as far as the organizational culture will allow it. Any organization that will succeed in a culture change must have some form of change management process." (*Maintenance and Reliability Best Practices*, p. 36)

A Real Example of Change

A friend was showing me his new phone recently at lunch, and he was particularly excited about it and the new features. He took some time to walk me through some of the capabilities and addressed how this new phone was superior to his old phone. With a very slight transition, we began to talk about change. As coincidental as it was, he was asking me about this very book you hold in your hand. I mentioned that I was piecing some thoughts together about change management. His response was classic, "People hate change." I responded that it seems the usual response to change was some sort of pushback, but I thought people accept change as a matter of life, they just don't like to be surprised by it for the simple reason that they don't have a reference on how to respond to it. (Author's note: I've just realized what a boring lunch partner I can be.)

I said, "Take your phone for example. You seem very excited about it." He told me he had waited a month to get it; it was a new company phone and he had been looking forward to getting the approval and then the phone. Couldn't wait to get to play with it.

I asked him if his attitude towards this new and upgraded phone would have been different if, when he got to work today, his boss had met him at the door and told him he had to switch to this new company phone, effective immediately.

My friend said, "Of course, that would have been incredibly inconvenient. I would have to transfer so much information."

If we are engaged in the *purpose* of the change, and have some input as to the *direction* we are intended to go, we will not be surprised by the need for change. I can't guarantee in every instance that change will be embraced and celebrated. But, as with my friend and his new phone, change can be an event that is anticipated for the better.

Getting Ready for Change

Here, Figure 1-3 is shown again for your reference.

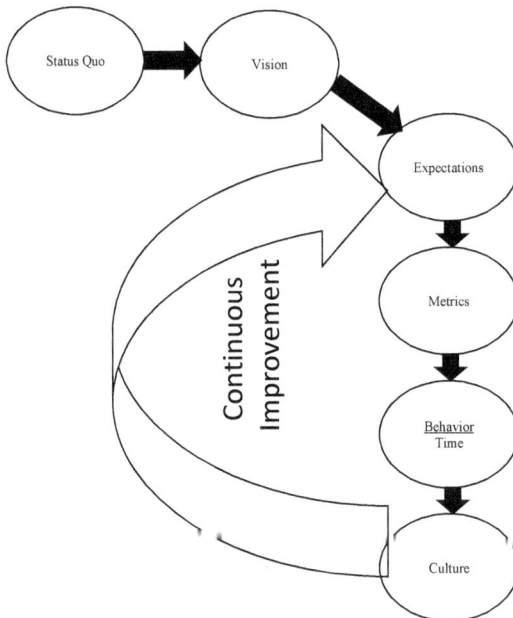

Figure 1-3 (reprinted): Continuous Improvement

There is a way to prepare an organization for change by easing or eliminating the shock.

Successful change management begins with the first elements of the model above. It is likely that the status quo and vision are poorly explained to the organization, if they are ever explained at all. For an approach to change that brings a bounty of positive effort, there are five significant elements that must be known and clearly communicated. This doesn't necessarily guarantee success because you still require leadership skills, but we certainly cannot be successful without these five:

1. Status Quo

2. Vision

3. Path

4. $$$ (money)

5. WIIFM (What's In It For Me)

Figure 2-16 shows a very simple model for change management, bringing the five listed elements into perspective.

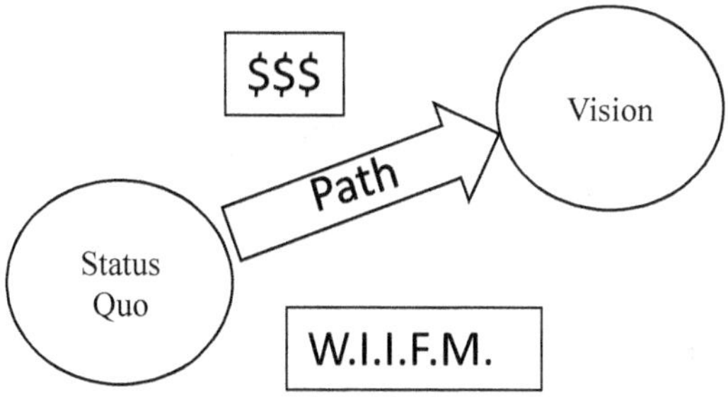

Figure 2-16: Simple change management model

Let's break down these five elements in greater detail.

Status Quo
Earlier, in Chapter 1, I mentioned that a colleague once told me, "Even if you know where you're going, you're still lost if you don't

know where you are." The event we were engaged in at the time was a maintenance and reliability assessment at a major food manufacturing plant. My workmate was the senior member of our team; however, I was assigned as the team lead for the evaluation. My partner was putting into words exactly the essence of what we were doing and came up with that phrase, which I now use almost every day. It's true, you know, that you can't get to where you want to go if you don't first establish clearly where you are now. And own it!

As it relates to change management, our first mission is to unambiguously state where we are at this moment. In the section prior to this, we talked about KPIs and metrics. We start our journey forward by establishing these values, for better or for worse. The key is not to be disappointed in this number, because it's an initial measure. Don't be disappointed, but don't be satisfied either. As another colleague stated and noted earlier in this text, "It's not the number you're after, it's the delta." I tell clients that this isn't the number we are stopping at, just a number we are passing through.

The organization, as a whole, will not move from the status quo until people become dissatisfied with the status quo. People need to be made to understand that the current state of things is not good for them, the organization, or the future existence of the enterprise. The status quo is not necessarily conveyed in overly emotional terms, but rather facts and logic. Pleading won't help, but a rational discussion, backed with data, can better set the tone for a needed change.

Ironically, the status quo isn't always a bad place for some people.

Note that threatening or making examples of people is the opposite of what is needed when making a case to move from the status quo. What's interesting about the current state of things in some locations, maybe even yours, is that regardless of how deficient the reliability and maintenance efforts are in bringing about high level reliability, some people in the maintenance organization 'have it made' just like it is. They will be very reluctant to change anything.

There are some maintenance technicians that can work all day and not actually do any work. When asked what they accomplished during the shift, they might say, "I tried to get on that machine five times, but the operator wouldn't let me have access. When I did get some time on a machine, the storeroom didn't have the part. I know, because I checked stores three times!" That person might 'have it made' exactly as things are right now.

Very important author's note: When defining the status quo, be brutally honest. Fair, but honest. When you're trying to lose weight, it serves no purpose to lie about your initial weigh in. Did you know that there is a little adjustment knob on the scale? It turns out you can weigh any weight you want to weigh. That doesn't make your pants fit better though.

Vision
Once the status quo is defined clearly and logically, with facts and data, the next element required for change management is the vision. We spent quite a bit of time discussing the vision earlier. In fact, if you've been participating in this workbook, you have actually created a draft vision statement.

The status quo and vision are not crafted in a vacuum. Each requires a small team, working together to put words to these vital elements. Consider a small group of maintenance and reliability folks, with ad hoc input from other agencies tasked to put these ideas together. If you can image that, you can start to picture the anticipation the group might have for the potential promises of the future, like my friend and his excitement for his new phone.

The path taken between the status quo and the vision is often rocky and filled with hairpin turns.

Path
The path is the route everyone in the organization must take to get from status quo to vision. If the route is fraught with obstacles and frustration, don't expect people to merrily go down the path. This particular aspect of change management is personal, very personal. So much so that I'd recommend sitting down with each person individually and discussing the path that they will take. People need to know what is expected of them. They need to know how they will fit in the new world order and have their fears and objections addressed.

My very first job as a civilian was as a plant engineer/mainte-nance manager for a manufacturing plant in rural Illinois. I was not prepared for civilian life and made a lot of what I'd like to call rookie mistakes. One notable offense was introducing computers to a maintenance organization that didn't even have a written work order system. In classic 'military-eze' I told the technicians that starting Monday, I wanted them to type everything into the computers. You can probably guess how successful that turned out to be.

I didn't take time to talk to the technicians individually to address and belay their fears and concerns. If I had, I would have discovered that 30% of my workforce couldn't read, and 100% of them couldn't type, nor had they ever touched a computer. It is hard to believe, but those days did exist. I don't make that mistake anymore.

$$$$

For any change to be accepted and fruitful, it has to make sense to and be supported by the people who are going to either profit from it or be the victim of it. The dollar signs and the W.I.I.F.M. symbols on the model give us the ammunition to make a compelling argument, up or down the chain of command.

Generally, we don't have to climb too far up the organizational ladder from the maintenance manager level before we reach levels that are occupied solely by MBAs or accountants. This group speaks a different language, and the discussions with these folks need to be preceded or quickly followed by a dollar sign, or several $$$. Every argument or quest for support must make sense to the bottom line. That's actually good for the organization, and thankfully, someone is paying attention to the fiscal sense of the enterprise.

W.I.I.F.M., Number One in Your Hearts, Number One on the Dial (Sounds Like a Radio Station)

The W.I.I.F.M. acronym simply means, What's In It For Me. Change discussions aimed toward those at levels lower than us in the organizational chart need to be focused on the value of the change for them personally.

Remember the computer story I shared earlier? I made a great $$$ argument to my boss, the VP for production. That's how I was

able to outfit my entire maintenance organization with computers (and dot-matrix printers). I didn't even think to make the
W.I.I.F.M. connection with my guys.

Let's see if we can piece this together, by using one simple
aspect of maintenance as an exercise.

Honestly state the current state, or status quo, of your preventive maintenance program. List the method, the results, the
efficiency and the effectiveness:

_____

_____

_____

_____

Clearly state your vision for the PM program:

_____

_____

_____

_____

Say you are in the elevator for twenty seconds with the CEO
of your company, and that person says, "Hey, I hear you want to
improve the PM program, what's that all about?" What is your $$$
response?

_____

_____

_____

_____

How do the associates in your maintenance organization
benefit from the new PM philosophy? What's In It For Me (them)?

_____

_____

_____

Finally, imagine that you are sitting down with every mainte-
nance person in your organization, individually, one-on-one, and
describing how they move from status quo to vision, profit from

it, and operate at this 'new' place. Name one employee in your
maintenance complex, and what you would say to that person
(make this personal):

Name:
_____

Message:
_____
_____
_____
_____

Look back at Figure 1-3, reprinted at the beginning of this
section. I ask you, how can any organization arrive at a new culture and then be propelled into organic and continuous improvement without an investment in the five important (and strategic)
elements of change management?

The phone scenario earlier has two possible outcomes. "How do
you like your new company phone?"

A. "I love it. It is going to help me do so many things."

B. "I hate it. I liked my old phone better and *I don't know why
they made me change.*"

The Communication Model

We have a problem, quite frankly. It's not that we have a lack of
communication, it's that we have lousy communication. There are
too many people talking and no one is listening. Prove that for
yourself. In the space provided, list all the forms of communication
that are currently used in your organization:

_____   _____
_____   _____
_____   _____
_____   _____
_____   _____

Here are some traditional forms of communication that have been noted around different industries:

- Face-to-face
- Email
- Text
- Other forms of electronic messaging
- Bulletin board posting
- Charts, graphs, tables posted

- Banners and posters
- Newsletters
- Message boards
- Scrolling TV screen messaging in break rooms
- Telephone

With all these modes and means of communicating, there are still people in the organization offering the excuse that "nobody told me." How is that even possible?

Thomas Peters and Robert Waterman shared the following in their best seller, *In Search of Excellence,* "The nature and uses of communication in the excellent companies are remarkably different from those of their nonexcellent [sic] peers. The excellent companies are a vast network of informal, open communications." (p. 121)

What it Takes to Really Communicate

First, to communicate, an organization has to have open channels for communication. Honest and open debate sends the message that, "We care enough to listen." Tom Rath and Barry Conchie suggest, "At a company-wide level, nothing creates stability as quickly as transparency." (*Strengths Based Leadership,* p. 88)

Apparently we haven't been communicating at all. We thought we were, but we weren't. Real communication requires a specific action and response. Most of us learned the elements needed for clear communication in high school. If not, Figure 2-17 might be a revelation.

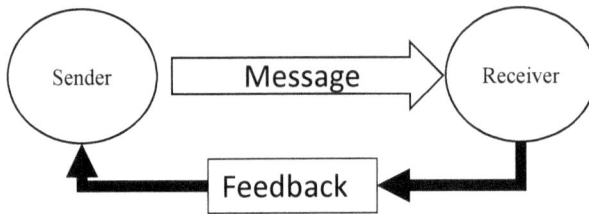

Figure 2-17: Communication model

Surprisingly, what is chiefly missing from most of our forms of communication is 'feedback.'

For all the forms of communication that you just entered in this workbook, go back and highlight those vehicles of communication which have a built-in, error proof feedback loop. I'm guessing you highlighted exactly one.

We know that we need feedback for effective communication.

Without the feedback loop we can never be sure that the intended receiver 'got' the message or understood the instructions. As lingering evidence of this fact we even add a post script to the end of our messages indicating that if anyone has any questions, "come see me."

The feedback loop is sometimes integral to life-significant information. Imagine any war movie or airplane cockpit exchange without the phrases: copy, roger, over, over-and-out, affirmative, copy-that, and "I repeat..." Those are professionals that understand the value of clear and understandable communication!

This assurance that the message was received and understood is woefully lacking in corporate communication today. The penalty for not responding to my electronic communication is that I send you another one. Finally I might call you on the phone. Before we would even get to the subject of our intended communication, I will ask, "I sent you an email, did you get it?"

When I was a small boy, my friends and I made the old soup can and a string phone. To this day I can't believe that this sort of device actually worked. Science is wonderful! Until we got the string nice and taut, we really couldn't hear one another. Our first message was always, "Can you hear me?" My friend would pull away from his soup can and yell directly to me "What?" I'd answer "I asked if you could hear me!"

Can your people hear you? Is the manner in which you are communicating effective? Do they understand what you want, is your vehicle of communication efficient, or does it result in confusion, missteps, and a phone call?

Example

Here is a very simple example, one that most readers can associate with.

The maintenance manager posts a simple notice on the bulletin board around 10:00 a.m. on the morning shift. Someone sees the note being posted and walks over to read it. They may actually get to the notice while the manager is still there, and the manager will actually stand there until they are done reading it and ask, "Does that make sense?" "Yep" is the only response.

The first technician to read it will notify others that there is a new notice on the board, and probably provide a preamble to the message, prior to the others actually reading it. The preamble is that technician's interpretation of what was written. If others on day shift have a question, they may ask the maintenance manager for clarification.

Around 3:00 p.m. the afternoon shift rolls in for work and starts the turnover process. Some may be put to work immediately if there is an issue that the day shift is working and immediate relief is needed. The afternoon shift may or may not get around to reading the notice until their first break, or their lunch break. They may have been alerted to a new message by others, but you never know. One thing is certain: their ability to ask the 'sender' a question during their shift disappears slowly as the clock inches toward the time the maintenance manager leaves for the day.

During break, there might be some discussion amongst the technicians on the afternoon shift about the notice, and some conclusion as to its exact meaning will be arrived at by consensus. Surely a uniformed interpretation will be offered to the midnight shift when they come in. Lack of clarity and certainty will be corrected by the source, or by the water-cooler, that much is certain.

The midnight shift is understaffed, and those on the shift have low seniority. They go to work keeping the plant running. There is a great likelihood that they will forget, or not bother to ask the maintenance manager for a better understanding of the message in the morning. And so the cycle continues.

People, agencies, shifts, and groups have different means in which they can be effectively communicated to or with. One common element exists in the need for effective communication, regardless of the mode... feedback.

In face-to-face communication, we aren't always asking for verbal feedback as to whether or not the audience heard and understood what was being communicated. In fact, often we are looking for the non-verbal clues. A head nod, the receiver writing down what's being said, or other signs that they 'get it.' Not to patronize, and I don't actually recommend this, but consider the times we gave explicit instruction to our children, "Now what did I just tell you?"

When we're providing information or asking for information, when we're communicating to our people, we're really asking "Now what did I just tell you?"

Just a quick note about committees and group sizes as they relate to clear and unambiguous communication. I have personally adopted the practice recommended in Jeff Sutherland's book, *Scrum,* and use it exclusively in my consulting work. He correctly observed that:

> "Once the teams grew larger than eight, they took dramatically longer to get things done. Groups made up of three to seven people required about 25 percent of the effort of groups of nine to twenty to get the same amount of work done. This result recurred over hundreds and hundreds of projects. That very large groups do *less* seems to be an ironclad rule of human nature." (p. 59)

Continuing:

> "So there's a hardwired limit to what our brain can hold at any one time. Which leads us back to Brooks. When he tried to figure out why adding more people to a project made it take longer he discovered two reasons. The first is the time it takes to bring people up to speed. As you'd expect, bringing a new person up to speed slows down everyone else. The second reason has to do

not only with how we think but, quite literally, with what our brains are capable of thinking. The number of communication channels increases dramatically with the number of people, and our brains just can't handle it." (p. 60)

EHS (Environmental, Health, Safety)

This section highlights what I believe is the most underrated, least understood, and most aggressively abused element in most organizations. With a high degree of confidence, I believe it would be safe to say that most organizations don't enjoy a cordial relationship between maintenance and safety. In fact, I've asked many times in places that I've worked, "What does the safety manager do all day?" To me it just seemed like they walked around the plant once a month and wrote fifty work orders. The visual from this was clear. Maintenance doesn't care about safety, if they did, there wouldn't be all these safety work orders! Of course we know that the safety manager and the safety department perform a valuable service. This extreme example was meant to illuminate the chasm that sometimes exists in our shared interests.

It doesn't have to be that way. In fact, if the foundation of understanding is established, and if the strategy clearly states the importance of these agencies working for the same outcome, a bountiful result could be in the offering.

Designing Process With Safety in Mind

I was working with a food company in the Midwest to develop a job kitting process between their maintenance and storeroom operations. The workflow was a smooth and classical design. There should have been no hiccups. I asked the simple question, "Is there a chance that a bin used for kitting parts could be contaminated by maintenance, and open up a risk for food safety, or for allergen transmission?" Of course there was.

The solution for this kitting process was to first understand that there was a risk, develop the countermeasures (with auditable elements), and present this to the food safety department for its scrutiny. There is a very good argument that could be made to involve the food safety associates in the kitting discussion at the beginning of the process development. Unfortunately, this question was asked well beyond that stage. The lesson learned was to note the need to incorporate the input from all those who may be affected by the work. Duly noted.

In no way am I suggesting that there be an adversarial relationship between the people responsible for maintenance and reliability and the EHS folks. I recommend and serve to connect these two branches of the same tree but we also cannot ignore the fact that there is often friction between the two.

We Are All On The Same Team

Sun Tzu is often credited with the phrase, "Keep your friends close and your enemies closer" in *The Art of War*, but that is not an exact quote. In *The Godfather II*, Michael Corleone shares similar advice from his father, Don Corleone, "Keep your friends close, but your enemies closer." The EHS folks are not the enemy, but we often have diametrically opposed missions. Ours should be work that is complementary and in the spirit of constancy of purpose.

"I don't care what it takes, keep it running" is a different leadership message than "Don't spill anything, get plenty of rest, and lift with your legs."

W. Edwards Deming brought us together with the thought of *constancy* and provided greater necessity, telling us it was vital to our survival as a company to work in partnership. "Create constancy of purpose toward improvement of product and service, with the aim to become competitive and to stay in business, and to provide jobs." (*Out of the Crisis,* p. 23)

Being diametrically opposed sometimes, even a lot of the time, is not, however, the same as being mutually exclusive. Our mission must be completed within the framework of understanding the requirements based on the known EHS risks. In order to work with EHS, we must know what the EHS goals are.

Safety Goals

In the space provided, list your organization's safety goals:

_____

_____

_____

_____

Some traditional safety goals might include:

- Incidents below the SIC (Standard Industrial Classification) Code

- Zero lost time accidents

- Successful results from each audit, internal or external

- Employee participation

- Culture change

Environmental Goals

In the space provided, list your organization's environmental goals:

_____

_____

_____

_____

Some traditional environmental goals might include:

- Zero regulatory infractions

- Reduction of environmental impact to the community at large

- Successful results from each audit, internal or external

- Employee participation

- Culture change

Planning Ahead

Granting an exclusion for the purpose of this discussion only, let's assume that an operation, a piece of equipment, or a process is designed correctly and constructed accordingly. Everything is as it should be, and environmentally, all aspects of the work have been addressed and confirmed to be correct. There always remains the possibility for error and catastrophe. In fact, the division between component failure and human failure is right down the middle.

In preparing contingency plans for environmental related possibilities, an organization has to take into consideration that anything can happen at any time. There will be a protracted discussion on Process Safety Management (PSM) later in this workbook, for now, let's consider that our role in reliability and maintenance is simply to design, install and maintain a machine or a process that is as inherently environmentally sound as possible. Our processes and our approach must be auditable and deficiencies must be noted and corrected immediately. All this documentation must be annotated in the work order system. That makes it searchable and provable.

Safety is Job 1!

In any organization safety and health must be the first priority.

The health and safety of our associates, our friends, is paramount to everything we do. There is so much that could go wrong. Just existing on this earth opens us up to an incalculable risk to our safety. Normal plant operations, with everyone performing safely, still expose people to an irrational level of near misses.

Going forward, we need to encourage—strike that—we need to *demand* the engagement of EHS in every aspect of equipment design, installation, operation, and maintenance. Having stated that, there also has to be some training and guidance to the environmental and safety folks. If not kept at some level of sanity these issues can actually do more harm than good for our companies.

I use an example when discussing scheduling maintenance work, specifically when addressing priorities. I draw a distinction between two extremes. "There is," I say, "a difference between a loose hand rail, and a missing hand rail." One is a priority 2 and one is a priority 1. This may be in direct conflict with good order and discipline at your location. If so, default to what is proper at your place of business.

In Table 2-1, you were asked to list the connection between certain roles and their place in the reliability effort. If maintenance and engineering were asked to make a connection between their roles and environmental, health, and safety efforts what would that look like? Use Table 2-5 to record your answers.

Everyone has a role to play in equipment reliability. Equally,

Table 2-5: Connection to EHS

Role	Connection to EHS
Maintenance Manager	
Maintenance Supervisor	
Maintenance Tech	
Storeroom Clerk	

everyone has a role to play in EHS matters. Our work is work to be accomplished together as partners. Reliable equipment is safe equipment.

Putting Safety Into A Proper Perspective

I was the plant engineer for a manufacturing plant in Kansas City for many years. The maintenance department worked for me, as well as the storeroom, and the planners/scheduler. It was a typical organizational structure.

One day I was overlooking the production floor from an elevated platform and the plant manager came up beside me and we began to talk. He said, "You know, we make three hundred parts a shift, that's our rate." I told him that I knew that. He went on to say, "An operator has a better day when we are making parts at a rate of three hundred per shift. In fact," he continued, "a maintenance technician has a better day at three hundred per shift as well."

The point he was making is that when everything runs as it should, no one is rushing around, reaching their hands into the machinery to 'un-jam' something. It's when we're dealing with a crisis that we start to make poor decisions that affect our health and safety.

"Very few people," he concluded, "reach their hands into a perfectly running machine."

That always stuck with me.

Chapter Summary

The business of reliability is a meaty topic, and an important element to gaining and sustaining momentum. We discussed some pretty heavy topics over the last several pages. If you've done your job, and I've done mine, we have some real structure to the direction the maintenance and reliability energy is going to be aimed.

Specifically, we discussed and worked on the ideas of:

- A vision

- A mission statement

- Goals

- Maintenance budget

- Metrics and KPIs

- Change management

- The communication model

- EHS and the importance of a partnership

Let's add to our reliability strategy by including this chapter's work with what has already been developed.

Our developing philosophy:

The goal of our reliability strategy is to decrease asset downtime and by doing so, increase equipment availability for our production partners. Our four prong strategic plan is to:

1. _____

2. _____

3. _____

4. _____

Our clear and focused vision is:

_____

_____

_____

_____

And our mission is direct:

_____

_____

_____

_____

We have three concrete reliability goals. Those goals are listed here, as well as the means intended to reach those goals, and the consequences expected as a result.

Goal	Means	Consequences
____	____	____
____	____	____
____	____	____
____	____	____
____	____	____

Goal	Means	Consequences
____	____	____
____	____	____
____	____	____
____	____	____
____	____	____

_____   _____   _____

_____   _____   _____

_____   _____   _____

_____   _____   _____

_____   _____   _____

Goal   Means   Consequences

With this vision, mission, and a strategy that supports our intended goals, we are confident that we can reduce the maintenance budget each year. In a controlled manner, working with the partners we value, we will reduce the maintenance spending by 3% per year.

_____

_____

_____

_____

Objective evidence of our progress will be tracked by the following metrics and KPIs:

The compelling argument to improve our current reliability efforts is a powerful message. The organization will benefit in a fiscal sense, meaning: ($$$)_____

_____

_____. The value to our workforce will be evidenced by: (WIIFM)_____

_____

_____

As we work together along this path of continuous improvement, we will communicate the progress of the individual activities by: _____

_____

Of specific value is the relationship between maintenance and operations, and maintenance and the Environmental, Health, and

Safety divisions. Our work will be to improve reliability while meeting the safety and environmental goals of:

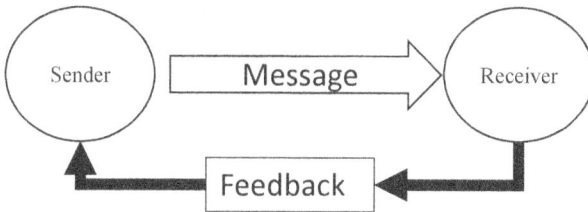

All this we will do while complying with the most basic communication needs as identified in the following diagram:

We will provide concise and complete communication as it relates to the reliability vision and anticipated deliverables. We will always seek and be open to feedback in an effort to engage the entire workforce and give everyone a 'seat at the table' when discussing equipment reliability.

THREE

The Process of Reliability

"The value of decisions depends upon the courage required to render them."

—Napoleon Hill, *Think and Grow Rich (p.142)*

If a person were to get lost in this maze of reliability excellence, this would be the place to do it. For some inexplicable reason, the idea of 'writing' something down seems abhorrent to us. Yet, without such documentation, we would truly lose the ability to *control* the reliability effort. I use 'control' in the most aggressive and extreme definition possible. Maintenance and reliability are things to be controlled; to be shaped and bent in our favor. The written word is the testament that, "at one time, we actually cared enough about it to write it down." As Napoleon Hill implies, the *value* in our decisions lies in the *courage* to make them. Is a decision not recorded really a courageous stance?

In a very basic sense, the difference between a decision and an opinion is that a decision is *written down*, recorded in some way. In that context, Napoleon Hill's words ring truer than ever.

What have we felt so convicted about that we took the trouble to document our actions?

I tell my classes that "if not for the first caveman painting on a cave wall how to kill a woolly mammoth, we'd all starve to death." Documenting the route to take for success saves us all the trouble of reinventing the wheel. It's these morsels of knowledge that form the record that we even existed in the first place.

In order for a *thought* to become an *idea*, and grow in complexity to form a policy or process, it must have roots buried within the organization. A grassroots effort can be born of an idea that the masses demand. If sincerely engaged, management need only to equip and resource the workforce, and then get out of the way.

In this chapter, we are going to develop the written word as it relates to reliability, and we are going to create the framework to build a very successful maintenance program. This information transcends manufacturing and is every bit as applicable to facilities management, engineering, and operations as it is to maintenance.

Ownership

A newer concept of 'ownership' was introduced in Chapter 1. A corresponding image appeared as Figure 1-5: What ownership actually means (reprinted here for your convenience).

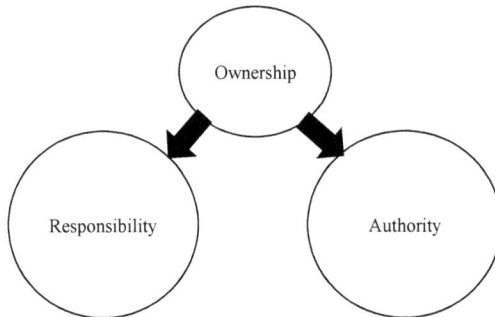

Figure 1-5 (reprinted): What ownership actually means

How does manufacturing process reliability square with the ideas formally presented regarding ownership? Surely the idea of bestowing authority on an hourly person runs in deep contrast to 'controlling' the maintenance and reliability work. Right? Wrong.

What is True Ownership?

Ownership is actually the key to sustainability, and it's with this knowledge that we can unlock the energy to develop processes and policies that actually 'stick.'

Recall that true ownership involves assigning some level of accountability and responsibility, along with abdicating a significant degree of authority. Generally, an issue that someone has authority over is something that they might care about. Here's a quick test of that theory: do you care for a rental car differently than you do your own car? When was the last time you washed a rental car?

Give it a Try
In the space provided, list one issue or 'thing' that you care about. Further, list how you have responsibility for it, and how you have authority over it. Here is an example:

Issue: My children's college education
Responsibility: Guide them through high school, encourage them
                  to attain good grades, help them select a college
Authority: I have all the money!

Now it's your turn:
    Issue: _____

    Responsibility: _____

    Authority: _____

When you provide an associate with the reins of ownership, that is, the responsibility and the authority, they will tend to feel a sense of ownership for the issue they now *own*. I've often said that if you ask operators how to squeak out 2% more on their production lines, they'll come up with some very creative and fiscally responsible ways to do just that. What's more impressive is that they will own the solution and work to make it a success.

To sincerely compel the workforce to adopt a policy or process developed to guide the organization forward, associates must sense a level of responsibility and authority about the policy or process. The most effective way to transmit this degree of ownership is to merely ask them for their inputs and opinions.

Surrendering any level of authority and empowering our associates with ownership requires a special kind of leader; a 21st century leader to be sure. Fortunately, we don't have to be a legendary leader to pull it off. "You don't have to be a leader of Lincoln's caliber to empower others. The main ingredient for empowering others is a high belief in people. If you believe in others, they will believe in themselves." (Dr. J. Maxwell, *The 21 Indispensable Qualities of a Leader,* p. 150)

Empowered employees who have great faith in themselves and a company hungry for fresh thinking is a formula ripe with great ideas.

An idea can often start as an opinion which leads to further action.

Your Opinions Are Your Ideas

Years ago, after a change-of-command ceremony, our squadron commander asked me to assemble all the officers in his office. Our commander had been the squadron commander for all of thirty minutes, yet he seemed to have something critical to say, for our ears only.

As the senior officer, I had been asked to tell my fellow captains and squadron lieutenants that the major wanted us in his office, ASAP. Once assembled, the new commander addressed us junior officers. "Gentlemen," he started, "here's how we're going to run this squadron while I'm in command. We're going to make decisions in this office. I'm going to ask for your opinion, and you're going to have one!"

The commander was telling us that, as officers in the squadron, we need to know what is going on. Have an idea what to do. Know the problems and concerns. Have an opinion.

Our new commander knew that once we were stretched and challenged to solve issues, his team would rise to the occasion and tackle anything. Once an associate, or a lieutenant, is let in on the inner circle of decision making and granted a level of 'authority' over an issue, there is no lengths to which they will not go to accomplish the mission.

Ownership of something you care about, and have responsibility and authority over can be a powerful motivator.

We've done this in practice, now let's do it for keeps.

Ownership in Practice

In the space provided, list one issue that is important to your reliability and maintenance effort. Assemble your maintenance team members and lay out the issue as you see it. Ask for their opinions. I think you'll find that your team is ready to tackle any mission with sound and plausible solutions.

Record their responsibility and the authority that you grant them.

Issue: _____

Responsibility: _____

Authority: _____

For comparison, here is an example from the reliability realm.

Issue: Completing PM tasks on time.
Responsibility: Understand your schedule, arrive on time, have the tools, parts and materials with you, complete the tasks according to the work order.
Authority: Demand an effective PM tasks listing, demand acces to the machine, demand spare parts and materials as needed, and demand enough time to do an exceptional job.

Look back on the example just shown. Does any of that seem unreasonable?

If there is a maintenance technician wielding that kind of authority, coupled with the responsibility, then that's a technician who owns that PM.

Commitment vs. Compliance

Granting the authority to make demands along with the responsibility to get the job done sets up the hope of achieving commitment vs. compliance. Make no mistake, each has its place. But in the long run, commitment will lead to continued excellence and total employee engagement.

There are many phrases that need to be banished from the business lexicon: *low-hanging fruit* and *synergy* are just two of them. More urgently, we need to eradicate the phrase *buy-in* from all our teaching and motivational materials.

It's the 'buying-in' that got us here! We need to stop now. Figure 3-1 makes this point clear.

Figure 3-1: Stop Buying-In   **BUY IN**

Here are a few action-paralyzing uses of the phrase 'buy-in:'

"We can't move forward until we get buy-in from the union."

"Just waiting for the boss to buy into our idea."

"Do you think the other folks will buy into what we're trying to do?"

A synonym for 'buy-in' is 'lip service.' We might think it means consensus or agreement, but it translates to apathy. Buy-in does not net commitment, it results in half-hearted compliance. No one really owns an idea that they bought into. In fact, buy-in indicates that someone had to be convinced of one method over another. True commitment is derived from the sense that an idea couldn't be successful without your input.

We don't want buy-in. What we want is understanding. We really want the people to understand the issue and make a determination as to how they can support it and benefit from it. When you truly understand something, you know what effect your actions will have.

Our commander, the same one mentioned earlier, told us at the end of the meeting in his office, "The decision we make in this office is the decision of this squadron."

It was clear to us young officers that we were free to talk amongst ourselves and the commander in that office. We could ask questions for understanding, and debate options and priorities. When the time came for a decision, we were going to reach one. Furthermore, we all understood the pros and cons and what we needed to do to support the decision.

The major didn't have to tell us what to do if we were asked by a subordinate, "Why are we doing this?" As an officer, you never said, "Because the major said so." That would be the death of an officer's authority. If it came down to it, if the sergeant wanted to know why a certain order had to be, I would say, "Because I said so."

Fortunately, our senior and junior Non-Commissioned Officers (NCOs) were superior in every way, and that flippant response was never needed. A good NCO always knew that an order was generally arrived at judiciously, and with good intent. Good officers always consult their NCOs about important decisions.

Commitment requires ownership; compliance does not. In fact, compliance mocks ownership in many ways. We often are required to be compliant with a subject that we have no ownership over. Sometimes we can be the victim of someone's zealous commitment; we have no ownership. This is an extreme example:

> In the 1500s, Hernando Cortez was the captain of eleven ships with more than 500 soldiers headed for Mexico to conquer the Aztecs and bring back gold and treasures. As you can well imagine, after his ships arrived in Mexico, the sailors and soldiers were not in the best of shape. Some of them fell ill on the journey, some had lost their motivation, and their quarters were not exactly shipshape. Several of Cortez's crewmates wondered what would happen to them in their strange new land. If they faced challenges or resistance, how would the crew return home? The captain had the perfect response: He burned the ships.
>
> 1. There was no going back
>
> 2. The only direction to go was forward
>
> 3. The old ways of doing things were about to be rethought
>
> 4. In fact, there were no more "old ways of doing things;" a new way had to be defined (M. Joel, *Six Pixels of Separation: Everyone is Connected, Connect Your Business to Everyone*, p. 200)

Fortunately, we will never find ourselves in a situation like this. We are, however, required to show a great level of commitment in our own right, and in our own era. Truly reflect on the following.

Are We Committed to Safety?

Virtually every company lists employees as its greatest asset. In fact, the safety of its workforce is paramount even to manufacturing the product.

List one safety compliance requirement that you have personal knowledge of that has been ignored or swept aside recently:

Hopefully, you listed something. You did if you were being honest.

I'd contend that if we had a true commitment to safety, there would be *zero* new work orders after a monthly safety walk-through of the facility. As is stands, any safety walk-through, of any facility, usually results in several safety work orders.

How can that be if we are totally committed to safety?

People are committed to what they own. What people always own is their opinion.

Ownership is Critical for Commitment

Our commander was smart, smarter than us junior officers anyway. He knew that we'd be more committed to an idea we helped to formulate and therefore, own, than to one that we simply were ordered to obey.

We want our people to be compliant with the federal, state, local, and company regulations and policies. That's not the kind of compliance we are discussing here in this workbook. We don't want our supervisors to execute a policy simply because we said so. We want our supervisors, and everyone in the organization to have an awareness of their roles in equipment reliability and maintainability. The most foundational means to creating the epiphany to awaken others to that responsibility is to invite them to engage in developing solutions for today's problems.

Engagement = Ownership = Commitment

Workflows

Everything we do is a process. Given that everything is a process, everything that we do can be mapped out. If everything can be mapped out, every single block of activity can be clearly defined and roles and responsibilities around that activity can be detailed.

Every task, action, and need can be mapped out and described in great detail. The workflow becomes our mechanism to engage others in how a 'thing' is supposed to work.

PBJ&B —A Tasty Example

Years ago, when my son was three or four, he asked if I would make him a peanut butter and jelly sandwich. Sensing this was my cue for immortality, I asked him if he'd like to have a dad's world-famous, triple decker peanut butter, jelly, and banana sandwich. With hopeful eyes that only a child could conjure up, he whispered, "Yes, please."

So began a decades-long tradition of the dad's world-famous, triple decker peanut butter, jelly, and banana sandwich. Sometimes it was served up cut diagonally, other times in two rectangles. It was never left as just one plain old square of a sandwich. The process was set. I was committed. I reigned supreme!

Years later, my son moved into an apartment right after college. His roommates were guys he went to college with. I drove down to his university town to check out his new accommodations. When I walked into the apartment, I said hello to his roommate, Alex, who was in the kitchen making a sandwich.

"Hi Alex, whatcha making?" I asked. His response, "A dad's world-famous, triple decker peanut butter, jelly, and banana sandwich." I wonder if he ever knew that I was *THE* dad. The process Alex followed was exact, in every detail. It's rumored that the recipe for this product of love still hangs in a tapestry over the greater Springfield, MO area to this day.

I wouldn't expect a PB&J sandwich to have a workflow associated with it. But if we were in the PB&J business, I'd insist that there be a workflow outlining the details necessary to ensure quality, safety, and value to the customer, along with making sure we weren't giving product away in product weight or waste.

A workflow simply defines the path necessary for a successful outcome.

A Workflow is a Map of a Process

There is a popular old criticism, "Sure, it looks good on paper." Of course it does! Who in their right mind would draft up something that looks bad on paper? A workflow defines the process it takes to make something correctly or perform a service correctly; essentially, converting that which looks good on paper to that which looks good in practice.

A process lights the path for a winning outcome. An outcome that is repeatable, and consistent. Again, recall a point made earlier regarding a company that was in the PB&J business. That business would have a workflow diagram defining the proper way to make a sandwich, right down to how much peanut butter, and what kind of jelly.

Let's continue by introducing a flow diagram. Figure 3-2 shows several of the more common flow diagram symbols and their meanings.

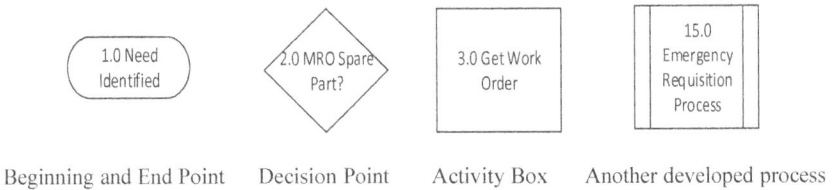

1.0 Need Identified	2.0 MRO Spare Part?	3.0 Get Work Order	15.0 Emergency Requisition Process
Beginning and End Point	Decision Point	Activity Box	Another developed process

Figure 3-2: Flow diagram symbology

Sketch out other flow diagram symbols you are most familiar with in Table 3-1.

Table 3-1: Other familiar flow diagram symbols

Meaning?	Meaning?	Meaning?

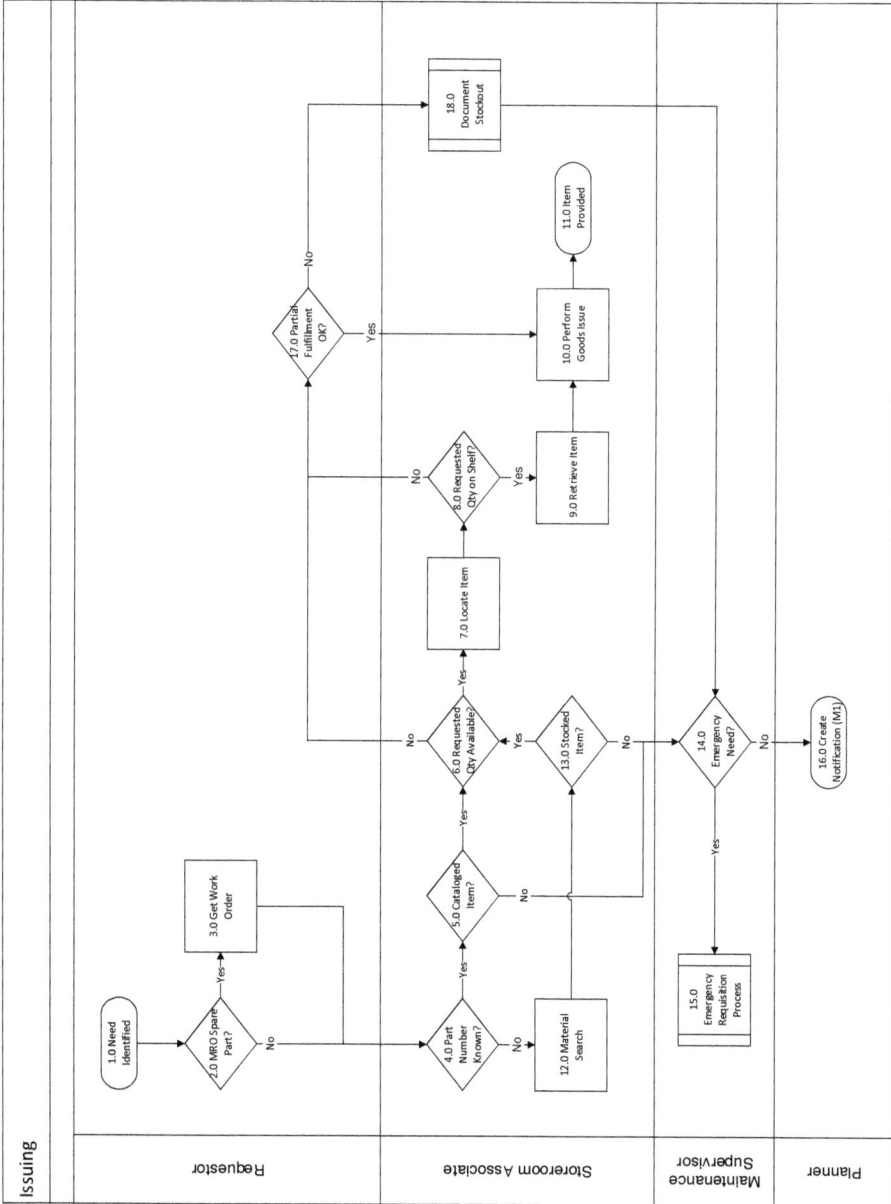

Figure 3-3: Sample workflow for issuing

Figure 3-3 is an example of a swim-lane style flow diagram.

Figure 3-3 is how the storeroom process of *Issuing* might appear if it was mapped out.

Do you have experience with flow charts and workflows?

In the space provided in Figure 3-4, draw one current process in your maintenance organization. Remember to number your blocks. There is no extra credit for neatness.

Figure 3-4: A current process

The workflow is a map that shows how work *flows* as the process moves forward. The workflow also helps to communicate who is responsible for the activities.

Identifying the Responsible Party

It would seem plausible to consider an eighth form of waste to amend our previous discussion on lean maintenance. It would be staggering to calculate the amount of time, energy, and lost opportunity that is trifled away each year because a clear line of responsibility is not known or generated.

The RACI chart and its merits are designed to bring this rogue management principle back to heel. RACI is an acronym for *Responsible, Accountable, Consult* and *Inform*. The RACI chart and the workflow complement each other and are companions to one another.

Note in the previous section, a workflow was introduced as an example of an activity that would normally be found in any maintenance organization; for this example, it was issuing a part from the storeroom.

The Workflow and RACI Chart Connection

If you hadn't noticed by now, each activity detail is numbered. The detail blocks are numbered, sequentially, to show the more desired and favorable route. Think of it as the critical path of a process. These block numbers and their descriptors are listed along the left side of the RACI chart, as shown in Figure 3-5 on the following page.

For each row in the RACI chart, there is one *R* denoting the responsible person, and one *A*, indicating the accountable person. For simplicity, the accountable person is usually the responsible person's boss.

There can be many *C*'s to highlight those who will be consulted on a certain activity; additionally, there can be many *I*'s to show that many people need to be informed of the process steps.

The true value of the RACI chart is often hidden in plain sight. We need to explore this telling document a little further.

Index	Activity	Reliability Engineer	Planner	Storeroom Supervisor	Storeroom Associate	Requestor's Supervisor	Requestor	Maintenance Manager	Maintenance Supervisor
1	Need Identified					A	R		
2	MRO Spare Part?					A	R		
3	Get Work Order					A	R		
4	Part Number Known			A	R				
5	Cataloged Item?			A	R				
6	Requested Qty Available?			A	R				
7	Locate Item			A	R				
8	Requested Qty on Shelf?			A	R				
9	Retrieve Item			A	R				
10	Perform Goods Issue			A	R				
11	Item Provided			A	R				
12	Material Search			A	R				
13	Stocked Item?			A	R				
14	Emergency Need?							A	R
15	Emergency Requisition Process								
16	Create Notification (M1)	A	R						
17	Partial Fulfillment OK?					A	R		
18	Document Stockout								

Figure 3-5: RACI Chart

Look at the RACI chart shown in Figure 3-5. The first R is under the Requestor for the task of Need Identified. The corresponding A along that same row appears under the title of Requestor's Supervisor. For this task, the Requestor is responsible for accomplishing the needed steps, and the Requestor's Supervisor is accountable for making sure it is done, and done correctly.

Consider this very subtle, but overlooked relationship. If the boss delegates the work to the subordinates, doesn't it stand to reason that the boss would equip and resource the subordinate to be successful? After all, bosses delegate most of their work to their subordinates. The subordinate's successful completion of all the tasks ensures that the boss' work will all be completed. It's the perfect order of things.

Also, with a clearly defined RACI chart, it is a quick task to develop a listing of roles and responsibilities for each function in the organization. Collect all the RACI charts that mentioned the position you are defining, and find the activities that have the R in their column. That is the stated and documented roles and responsibilities for that position.

The RACI chart also serves as a real-time needs listing for training back-up support, and developing succession plans. Again, reference Figure 3-5. If the Planner were on vacation when it was time to 'create a notification,' we could train another person on the task that the Planner is responsible for.

Consult and Inform are frequently a forgotten duo when compiling a RACI chart.

Consult actually has two connotations. Consult clearly indicates that the person or persons which have been identified with the C are to be consulted with on the meaning and activity detail around that specific step. Consult, however, has the additional meaning of identifying those who should be consulted with if there is a conflict with the interpretation of the detail steps, or if the activity results are trending in a negative direction.

An Example of Internal Consulting

An example might include a trend of having too many parts requests made at the storeroom without a work order. Are there some people at your location who could figure out a method or two that would keep someone from showing up at the issue window, for a part, without a work order? The people that you might be thinking of would be the ideal candidates to serve as consultants on this particular step.

A RACI Chart Using Your Example

In Figure 3-4, you created a simple workflow. Refer back to your simple workflow. In Table 3-2, create the corresponding RACI chart.

This RACI chart will be used in the following section where we'll discuss and build a process guide.

Table 3-2: Draft RACI chart

Index Number and Activity (list this information in the rows)					

Process Guides

Everything we do is a process. As such, what we do can be mapped out. The process, as we discovered earlier, guarantees that we have a predictable outcome. Preferably, the outcome will be the desired outcome. It looks good on paper so it should look good in practice.

Definition

A process guide is a collection of aspects about a particular process. The primary portions of a process guide include, but are not limited to:

- Business and Process Description
- Applicability
- Business Process Overview
    - Inputs
    - Outputs
    - Touchpoints
    - Key design decisions, requirements, and expectations
- Business Process Workflow
- RACI chart
- Activity detail

The *business and process description* is a summary of what the process guide covers. Specifically, it lays out the argument as to why there needs to be a documented, step-by-step procedure and compels others to take notice.

Not all process guides are pertinent to all parties in an organization. The segment regarding *applicability* is tuned to 'speak' to those whom the process guide is aimed to guide. The applicability section also provides a clear indication as to who owns the process and what agencies were engaged in the development.

The *business process overview* covers some of the more critical elements of the process.

Inputs list the necessary, and often times absolutely required information, data, forms, and decisions. *Outputs* include all the approved finishing points of a process.

Touchpoints provide notice to those navigating the process guide as to what other processes are invoked during the process completion. A touchpoint is simply a point at which the process being executed 'touches' another process.

Key design decisions, requirements, and expectations is the area in a process guide that details all the steps and actions that have to be performed correctly for a process to be successful. This is a quick check of the most critical steps in the defined process.

Workflow and *RACI Chart* have recently been defined.

The *activity detail* could be considered the main body of the process guide. It is certainly the portion of the guide that has the most specific information. The activity detail mirrors the workflow and RACI chart in activity titles and sequencing. Additionally, the activity detail steps explain what action is to be taken for a singular step, who is responsible for it, and what the next required step is to be.

There are nine significant maintenance management processes:

1. Identify	6. Turnover Execute
2. Approve	7. Return to Service
3. Prioritize	8. Closeout
4. Plan	9. Evaluate
5. Schedule	

There are additional maintenance reliability processes, four of which are:

- PM/PdM
- Work request creation
- Work order creation
- Job plan creation

There are twenty-eight primary storeroom processes:

1. ABC Classification Review	15. Ordering
2. Adjust ABC Classification Model	16. Preservation Program
3. Add to Stock	17. Receiving
4. Bench Stock	18. Repair or Replace
5. Critical Spare Parts Algorithm	19. Return to Stock
6. Critical Spares Review	20. Return to Supplier
7. Cycle Counting Criteria	21. Salvage Value
8. Cycle Counting	22. Special Tools
9. Data Scrubbing	23. Standardization
10. Disposal of Scrap	24. VMI/Consignment
11. Document Stockout	25. Item Substitution
12. Emergency Procurement	26. Kitting
13. Issuing	27. Obsolescence Criteria
14. Optimizing Stock Levels	

What follows over the next few pages is an example of a process guide. This example is for the storeroom process of 'Issuing.'

A Real Process Guide—as an Example

The example that follows is an offering of an actual process guide developed for a client. If your organization has a formal method for composing processes and process guides, please use it and get very good at it.

Issuing

Business Process Description

The purpose of the Issuing process is to set forth guidelines for issuing items from the Storeroom. All items issued from the Storeroom inventory must be issued to a valid work order and tracked against the work order's financial authority for accountability. Item issues will only be made directly to a valid work order with or without a reservation for the item and will not be made directly to a cost center.

The primary roles involved in this process are:

- The Requestor – make s request for item(s), take s possession of item(s)

- The Storeroom Associate – confirms the presence of a valid work order, retrieves item(s), documents stock ut occurrences, issues item(s)

- The Maintenance Supervisor – determines if the need is an emergency, ee cutes the Emergency Requisition process

- The Planner – creates a notification

Applicability

The Issuing process is owned by the Maintenance Department with consultation from Production as the primary end users. The target audience for distribution of this process includes, but is not limited to, the following departments: Storeroom, Production, and Procurement.

This process is not intended to be used for issuing items for capital projects.

This process is to be used for the issuance of all items issued, tracked, and controlled by the Storeroom to include parts, materials, and consumables related to plant Maintenance and Production. This process is not intended to be used for tools or everyday consumables utilized within the administrative or office setting (e.g. office supplies).

Business Process Overview

This section highlights the necessary inputs, desired outputs, and associated touchpoints (where the Issuing process relies on or interfaces with other processes), as well as outlining the key design decisions, requirements, and expectations.

The following **input** is necessary to complete the Issuing process:

- Valid work order

The following **outputs** are the expected result of the Issuing process:

- Item issued to the Requestor

- Adjustment of on-hand inventory to reflect issuance of item(s)

- Notification that the item is not available

- Work notification, M1

The Issuing process has the following **touchpoints** with other processes:

- Execution of the Issuing process could involve the execution of the Document Stock out process

- Execution of the Issuing process could involve the execution of the Emergency Requisition process

The following are the **key design decisions, requirements, and expectations** for the completion of this process:

- Proper execution of the Issuing process aids in keeping inventory accurate, and helps when researching root cause issues (for history)

- No items are issued without a valid work order

- Items that are to be issued under this process must first be items that are in fact stocked items to begin with

Business Process Workflow

The following flow chart gives a high level overview of the process flow for Issuing. It illustrates necessary inputs, desired outputs, and any touchpoints (where the Issuing process relies on or interfaces with other processes)

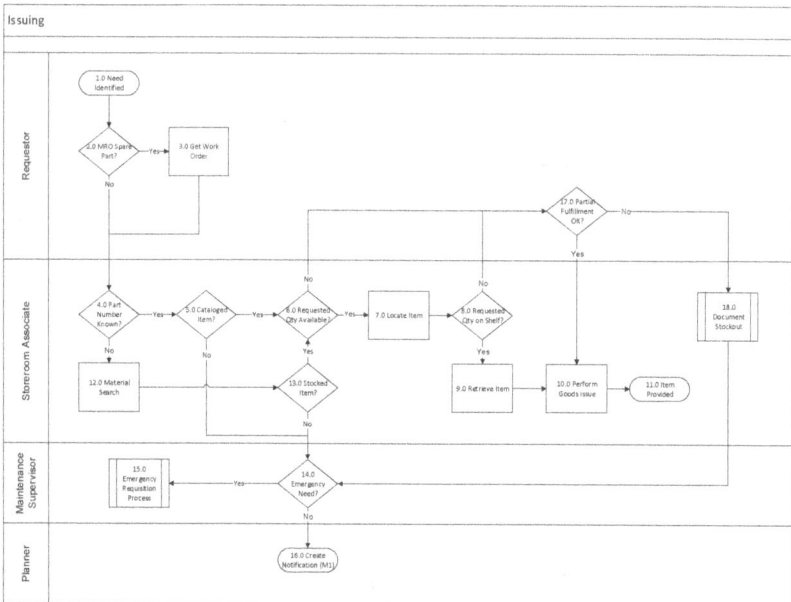

RACI

The RACI (Responsible, Accountable, Consult, and Inform) chart indicates the type and level of involvement required from designated roles in the organia tion for each activity of the Issuing process.

Activity Detail

The following section details the activities necessary to complete the Issuing process. This ep ands the information contained in the workflow.

1. Need Identified

The issuing process begins when the Requestor identifies that they have a need for a particular item

Activity completeG o to activity 2, MRO Spare Part?

Index	Activity	Reliability Engineer	Planner	Storeroom Supervisor	Storeroom Associate	Requestor's Supervisor	Requestor	Maintenance Manager	Maintenance Supervisor
1	Need Identified					A	R		
2	MRO Spare Part?					A	R		
3	Get Work Order					A	R		
4	Part Number Known			A	R				
5	Cataloged Item?			A	R				
6	Requested Qty Available?			A	R				
7	Locate Item			A	R				
8	Requested Qty on Shelf?			A	R				
9	Retrieve Item			A	R				
10	Perform Goods Issue			A	R				
11	Item Provided			A	R				
12	Material Search			A	R				
13	Stocked Item			A	R				
14	Emergency Need?							A	R
15	Emergency Requisition Process								
16	Create Notification (M1)	A	R						
17	Partial Fulfillment OK?					A	R		
18	Document Stockout								

2. MRO Spare Part?

The Requestor, possibly in consultation with others, will determine if the needed item is an MRO spare part. A spare part requires a work order so the withdrawal from stores can be charged to an asset.

> Yes – Go to activity 3, Get Work Order
>
> No – Go to activity 4, Part Number Known?

3. Get Work Order

Requestor must obtain a work order to be presented to the Storeroom. Preferably, the valid work order will list a part number to assist the Storeroom Associate in locating the required item. In the event the Requestor does not have a valid work order it is the responsibility of the Requestor to obtain one before the Storeroom Associate can issue any item(s).

> Activity complete – Go to activity 4, Part Number Known?

4. Part Number Known?

The Storeroom Associate will determine if the item being requested has a known part number. A part number could indicate that the item is stocked, or at the very least, can be ordered to specification. Otherwise, additional steps are necessary to locate the item.

> Yes – Go to activity 5, Cataloged Item?
>
> No – Go to activity 12, Material Search

5. Cataloged Item?

The Storeroom Associate will determine if the item being requested is a cataloged item. A cataloged item is one that has a known part number, and all associated information to order, but may not actually be an on-hand stocked item.

> Yes – Go to activity 6, Requested Quantity Available?
>
> No – Go to activity 14, Emergency Need?

6. Requested Quantity Available?

When attempting to issue an item to a work order, the Storeroom Associate will validate if there is sufficient quantity on hand to satisfy the request via the inventory management syst em.

In the event there is sufficient quantity showing on hand, the MRO Storeroom Associate will obtain the item and issue it.

In the event there is insufficient quantity showing on hand,the MRO Storeroom Associate will inform the Requestor of the potential problem with immediately filling the request.

In the event there is an on-hand quantity but it is insufficient to satisfy the request, the MRO Storeroom Associate will offer to partially fill the request, but leave the decision to do so up to the Requestor. If the Requestor chooses to take the partial issuance, the Storeroom Associate will issue the on-hand inventory.

> Yes – Go to activity 7, Locate Item
>
> No – Go to activity 17, Partial Request Fulfillment OK?

7. Locate Item

The Storeroom Associate will locate the item being requested.

> Activity complete – Go to activity 8, Requested Quantity on Shelf?

8. Requested Qty on Shelf?

In the event that the item or a sub-set of the quantity is not the amount needed (that it might be different than what was initially listed in the inventory syst em previously , the Storeroom Associate will inform the Requestor.

> Yes – Go to activity 9, Retrieve Item
>
> No – Go to activity 17, Partial Request Fulfillment OK?

9. Retrieve Item

The Storeroom Associate will retrieve the item to be issued from the location shown in the inventory management syst em.

Activity complete – Go to activity 10, Perform Goods Issue

10. Perform Goods Issue

The Storeroom Associate will record the item's removal from stock in the inventory syst em.

Activity complete – Go to activity 11, Item Provided

11. Item Provided

The Storeroom Associate will provide the item to the requestor.

Activity complete – this completes the Issuing process

12. Material Search

In the event the part number is not kn own, the Storeroom Associate and the Requestor will work to locate the item in the inventory syst em or by other available means.

Activity complete – Go to activity 13, Stocked Item?

13. Stocked Item?

If located, or not, it is the responsibility of the Storeroom Associate to determine if the item being requested is even a cataloged item, and within that description, if it is a stocke d item (indicating that an inventory level is exp ected).

Yes – Go to activity 6, Requested Qty Available?

No – Go to activity 14, Emergency Need?

14. Emergency Need?

The Maintenance Supervisor, in consultation with the Requestor, will determine if the need for the item is an emergency.

Yes – Go to activity 15, Emergency Requisition process

No – Go to activity 16, Create Notification

15. Emergency Requisition Process

The Emergency Requisition Process is a separate process; please see the relevant documentation for actions required.

Activity complete – this ends the Issuing process

16. Create Notification

The Planner will create a notification for M1 jobs, listing the needed items.

Activity complete – this ends the Issuing process

17. Partial Fulfillment OK?

In the event there is an on-hand quantity but it is insufficient to satisfy the request, the Storeroom Associate will offer to partially fill the request, but leave the decision to do so up to the Requestor. If the Requestor chooses to take the partial issuance, the Storeroom Associate will issue the on-hand inventory.

Yes – Go to activity 10, Perform Goods Issue

No – Go to activity 18, Document Stockout

18. Document Stockout

Document "Stock ut" is a separate process; please see the relevant documentation for actions required.

Activity complete – Go to activity 14, Emergency Need?

This completes the Issuing process.

Work on Your Process Guide Element Skills

Earlier you were asked to create a simple workflow and a corresponding RACI chart. From those exercises, now it is time to put together a process guide following the format of the sample just shared.

Don't panic, we are going to walk through this together. What was the process you chose to map out? _____

_____

What would be the business process description?

_____
_____
_____
_____
_____

Who would be involved and what are their roles?

_____
_____
_____
_____

Who would own this process, and who would it be applicable to?

_____
_____
_____
_____
_____

Complete the process overview:
Inputs & Outputs:

_____
_____
_____
_____

Key Design Decisions, Requirements, and Expectations:

_____

_____

_____

_____

You've already built an effective business process workflow and
RACI chart through these past few exercises. The last step is to
define each activity. In the space provided, detail out just three
activities. We will use these to build your skill at detailing activities.

Activity detail #1:

_____

_____

_____

_____

Activity detail #2:

_____

_____

_____

_____

Activity detail #3:

_____

_____

_____

_____

The work just accomplished: the workflow, RACI chart, and process guide are prime resources that you may now use to assemble a
concise and exceptionally detailed document outlining the roles and
responsibilities of stated positions for a specific process.

Roles and Responsibilities

For some perplexing reason, creating a document defining roles and responsibilities seems to be quite difficult in most organizations. The result of this indecisiveness is placing people in positions that aren't neatly defined, requiring them to perform work they are not suited for or trained to do, and may have a negative impact on morale and general contentment. To be honest, how many people are performing the exact job that they understood the job to be?

A Major Stumbling Block to R's & R's

One primary detractor from good order and discipline in developing roles and responsibilities is the mere fact that we often think of *roles* and *responsibilities* as synonyms. The have very different meanings. The role, the specific role, actually aids in laying out the particular responsibilities of that role.

Here is a quick example to demonstrate the difference between roles and responsibilities.

A maintenance planner/scheduler may have ten or more very distinct roles, for example:

- Planner
- Scheduler
- Facilitator
- Maintain Bills of Material (BOMs)
- CMMS gatekeeper
- PM/PdM overseer
- Picklist generator
- Data analyzer
- Report generator
- KPI/metric developer

Every one of these distinct roles has its own set of precise responsibilities. For example:

As facilitator, the planner/scheduler will:

- Ensure maintenance technicians are engaged in job plan creation

- Ensure maintenance technicians are engaged in job plan feedback

- Act as primary facilitator of the weekly scheduling meeting

- Perform duties as facilitator during Preventive Maintenance Optimization workshops

Continuing with the workflow, RACI chart, and process guide you've been creating, it is now time to extract some roles and responsibilities for one agent in your example.

In Table 3-3, refer to your process guide and supporting information recently completed to identify one position to use as an example. List that position's roles for that specific process, and list their responsibilities as they relate to the process. Hint: if you are ever stumped as to what roles and responsibilities to list, just reference the RACI chart.

Table 3-3: List of Roles and Responsibilities

Position	Roles	Responsibilities

Congratulations! You have just successfully constructed a process guide and determined the roles and responsibilities to make the process consistently successful. If you complete this task for real, not as an example, but for real, for every process, you may be in the top 5% of leading reliability practitioners in the world. A very conservative 95% of maintenance and reliability efforts fail because the leader never writes anything down. You can be, and you are, better than that.

Continuous Improvement

A segment on continuous improvement might seem disjointed at this juncture and a little out of place. After all, we just completed a series of discussions leading to an effective process creation. How does continuous improvement logically follow that?

For an answer, let's review Figure 1-3 once more, remembering that this figure indicates how continuous improvement works with and within a basic culture change model.

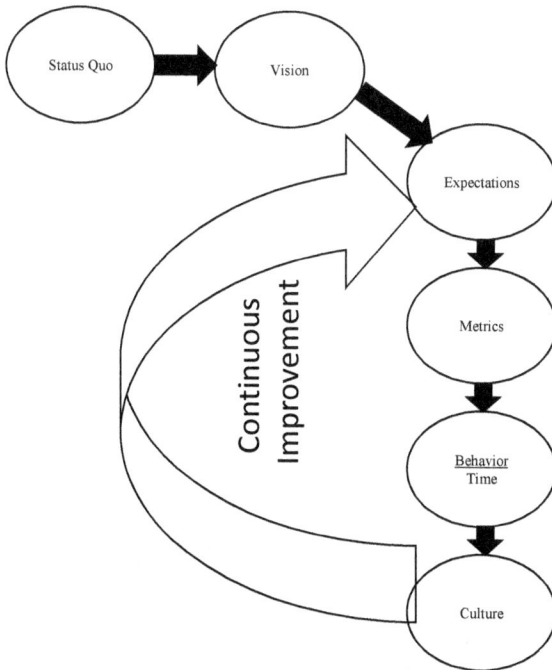

Figure 1-3 (reprinted): Continuous improvement

In Figure 1-3, the oval titled *Vision* is meant to articulate the future state. The leader builds and communicates the vision, not alone, but with engaged associates focused on improving equipment reliability. The process guide is the vision in the tangible form. Not one process guide, of course, but all the process guides collectively make up the vision of what is to be. As I'm fond of saying (and being quoted), "Not Monday, but a year from Monday."

The battlefield of reliability is littered with leaders who made the potentially fatal tactical error of moving from status quo, or even vision, directly to *culture*. This is a major blunder, not always fatal, but often crippling. I know; I made that mistake myself.

What is it that we are continuously improving?

What's Really at Stake Here?

Improvement can only happen to something that is growing and evolving. A rock cannot improve. It will always be a rock. A plant, on the other hand, can improve from a seedling to a vibrant adult plant.

In Figure 1-3, notice again how the continuous improvement takeoff is from the culture position. By way of the process guide (vision), expectations (roles) and metrics (responsibilities) are refined over time to lead to a new sense of culture. It is only during this dynamic portion of momentum that true improvement can be suggested. Everyone has to be *moving* to be moving forward.

The consistent and proper execution of the process is what fuels continuous improvement.

For example, each diamond shape in a workflow represents a decision point. There can be up to three outcomes from a decision diamond. Typically, one decision is the most favorable. An organization should work to improve the results of that particular decision diamond and improve the conditions until an overwhelming ratio of decisions are in the positive direction.

This approach not only identifies the areas to be worked on, but gives everyone a clear idea of what direction will lead to greater effectiveness for that process.

Here is a classic example of the marriage between a process and continuous improvement.

This May Sound Familiar

In the storeroom process of 'Issuing,' there may be a decision diamond that asks, "Does the requestor have a work order?" There are reasonably two responses to this question: yes or no.

This is a very easy metric to calculate, and it is most beneficial if the overwhelming majority of responses are 'yes.'

Listed in the storeroom process guide for 'Issuing' are all the steps necessary to make sure that technicians have work orders when they request a part from the storeroom. In the guide are the roles and responsibilities, with incredible detail, and an insistence on having a work order when requesting a part. However, many times a part is requested and no work order is presented. How can this be?

Establish a Baseline

Continuous Improvement (CI) always starts by marking where the journey begins. In our case, in the example above, let's say a work order is presented 27% of the time. This was measured over a period of six weeks to provide some statistical relevance. CI would dictate that we work in cooperation between maintenance and the storeroom to improve this result. We are not going to bring it up to 100% overnight, but incrementally; from 27% to 35% to 40% and so on. Remember the Darwin quote at the beginning of this workbook?

Without a written and communicated process, there would be no agreement as to what precise elements of an activity need to be improved, or honed. There would be a cloudy understanding of who should mandate the improvement and in what direction the team should be heading. If there is no written and understood means to accomplish a task, or level of competence acceptable in an activity, what exactly is the value of attempting to improve it? It's working exactly as defined now.

If You Can't See It, You Can't Be It

Years ago, I was consulting with a maintenance manager at a facility in the Northwest. He mentioned that he wanted his maintenance folks to do a better job of getting parts out of the storeroom. I asked him how they were supposed to do it. He proceeded to tell me until I stopped him at the beginning of his explanation. "Don't tell me how they're supposed to do it." I said. "Show me the process they are supposed to follow."

My client told me that there was no written process. "Seems to me," I remarked, "that they are doing it now exactly as they understand it should be done."

Having no clear direction or path to take can lead to disorganized efforts to comply at best, and chaos at worst. You can't build on chaos.

Our evolution, our continuous improvement, should be done slowly and deliberately. It should be done slowly because we are moving bodies, minds, talents, concerns, egos, and history. It should be done deliberately because everyone deserves to be 'in on the plan' and should feel engaged and knowledgeable about what is expected.

The improvement term *Kaizen* has been loosely translated to mean 'just a little bit better today than yesterday.' I think that's a good way to improve. Improving, incrementally, is a practice proven in nature. While we're growing with purpose, we might as well grow for the right reasons. Deming said, "Much of American industry is run on the quarterly dividend. It is better to protect investment by working continually toward improvement of processes and of product and service… " (*Out of the Crisis*, p. 98)

Once an area of concern is recognized and determined to be a target of an improvement effort, an organization should determine the proper tools necessary to bring about a correction.

RCA—Problem Solving

A common misquote is that most humans only use 10% of their brain. Actually, we use 100% of our brain, but only 10% of our brain's capability. The idea of using only a fraction of something powerful transitions nicely to the topic of *Root Cause Analysis* (RCA), or problem solving.

It is very common for organizations to use RCA techniques primarily for safety incident investigations. Most likely, an organization will have an RCA process for equipment breakdowns, but it may take on a punitive feel rather than a positive message.

The truth is we've been going about problem solving in the wrong way, and when we are forced to conduct some type of Root Cause Analysis, this error in execution, coupled with a negative connotation, nets us a less than desirable investigation. We put in all the energy, but usually gain nothing of significance as a result.

Develop a Common Understanding

It might be helpful to agree on a few terms to start. In the space provided, write your definition for the phrases given. Don't write the definition you think I want, actually write what you think it means:

Root Cause Analysis:

_____

_____

_____

_____

Problem Solving:

_____

_____

_____

_____

Definitions

Root Cause Analysis refers to the practice of analyzing data and information to arrive at the primary cause(s) for an unwanted event or trend. This is a street definition. You are likely to come across more formal explanations in other reference material. This, however, will work for our needs.

Problem Solving is the act of defining a problem and its consequences and creating at least one, but probably several, options to correct the issue. Again, this is a street definition, but it will work for our discussion.

Although this section is titled RCA—Problem Solving, these two topics are vastly different. It would not be unusual for you to have defined these two terms to mean very similar things. Breaking down the definitions will help to understand how they are different, how they can and should be used differently, and will shed some light on their areas of similarity.

What's the Difference?

For Root Cause Analysis, a requirement was listed that it must be tied to an unwanted event or trend. We don't have to wait for something to fail, or break completely before conducting an RCA study.

Consider these two examples:

1. A primary production machine fails, resulting in two hours of lost manufacturing.
2. An associate slips in the parking lot on black ice while walking to the plant.

The first example is fairly common. It would be typical for a manufacturing facility to establish some sort of trigger point that necessitates the execution of an RCA. In the second example, there might not be any action unless the associate was injured, or they formally commented on the dangerous situation in the parking lot.

One example could result in a Root Cause Analysis, the other in a problem solving session.

Any study and conclusion coming from an RCA related to Example One would actually be the result of a Root Cause Failure Analysis because the failure has already happened. An RCA is conducted when something is *trending* in an unfavorable direction, or an unwanted event has occurred. Of course, the breakdown is an unwanted event, but the differences in RCA and RCFA are subtle and a bit more nuanced.

Let's assume for Example One that there was no trending, so we don't have to focus our attention on that portion of the definition. Rather, we had a component fail on the machine. The RCFA would lead us to the offending part, perhaps rather quickly This would do for completing the paperwork, and reporting some conclusions up the chain of command. But the analysis doesn't stop there.

A very critical element in the definition of RCA is that it is an analysis of data and information. Most organizations can conclude in very short order that they have horrible data and information. The bare minimum has likely been collected over time on maintenance work orders.

In the first example given, the result of the failure was the loss of two hours of manufacturing time. This is most certainly an unwanted event, as we've already concluded. Our RCFA study has uncovered that the offending component leading to the breakdown was a bearing. A simple, commonly stocked bearing has cost the company perhaps thousands of dollars.

In this instance, it is almost certain that the plant or facility manager is going to ask, "Who PM'd that last?" We are going to discover in Chapter 6 that the real question ought to be, "What's the failure mode and what are we doing about it?"

How to Leverage the Differences

RCA now comes into play when we determine that the bearing should have lasted about 5,000 hours, but only lasted 3,800 hours. This was a premature failure, for some reason (cause or causes), and as a result, many dollars of revenue were lost.

I keep mentioning cause or causes. The result of an RCA study can be more than one primary cause. As it is with our bearing in this example, we know that a bearing can fail for a number of reasons, or causes. Any one or multiple causes could be true in this example:

- Lack of proper lubrication

- Misalignment

- Stored in the storeroom too long; the grease has a shelf life

- Stored in the storeroom too long; the bearings brinelled

- Improperly stored in the storeroom; not protected by wax paper and box

- Improperly stored in the storeroom; shelf had constant harmonic vibration

- Improperly installed

- Improper operation

- Purchasing decided to buy a cheaper bearing

- Operator abuse

Determining the root cause or causes from this list will require extremely good documentation of maintenance and storeroom execution.

A few comments on when to conduct an RCA study: quotas are perhaps the worst way to identify the need for a formal Root Cause Analysis. As a bit of a contradiction, there should, however, be trigger points that give guidance on when an RCA needs to be conducted.

What is most desirable is for the team to determine that the best means to identify the underlying culprit to any unwanted event is to conduct some formal study of the facts. Quotas tend to net a rush to judgement and will often lead to very shoddy work and a sense that the whole effort was done with little enthusiasm.

A major error in leadership as it relates to RCA is believing that members of the workforce won't take it upon themselves to formally study and solve major upsets in production or service. Another mistake is the belief that if management doesn't dictate the terms (or triggers) of when to perform an RCA, and what actual RCA tools to use, the workforce members will be apathetic in selecting any for themselves.

This notion flies in the face of the ownership model we discussed in an earlier section. Of course, there should be some structure around formal RCAs, but there also needs to be some line of responsibility and authority with the direct labor force. A balance would be beneficial.

There are many formal tools to use for Root Cause Analysis. Most of them work well, and are easy to navigate. Pick one or two of these tools and get really good at their execution. Some of the more common tools are:

- 5 Why

- Cause-and-Effect

- Fishbone

What RCA tool do you use?

_____

_____

For any tool, never underestimate the power of creating a drawing around the area of concern. Certainly, walk down the area, but for the analysis, actually sketch the machinery. Engineering drawings and blueprints are helpful, but I find that the mere act of drawing out the equipment generates discussion, debate, and consensus. This is a very powerful concept.

A Real-Life Example

Figure 3-6 is an example of when such a practice proved very helpful. This was a Root Cause Analysis conducted on a hydraulic coupling that failed on a primary water pump. The original drawing was sketched on a dry-erase board in the Middle East. I used Microsoft Excel to create the drawing on my computer for mobility of the information.

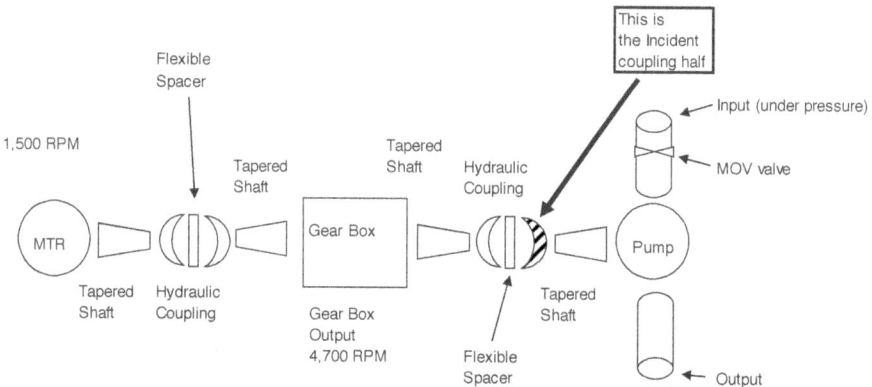

Figure 3-6: Sketch of water pump with a hydraulic coupling

This particular organization used a cause-and-effect tool to arrive at a root cause. The trend that was attempting to be solved (it was also an unwanted event) was that the pump side coupling hub kept spinning on the pump shaft when the asset was started up. Since this was a hydraulic coupling (the only one this author has ever seen), there was no key, spline, or set screw holding the hub to the pump shaft. The mechanism that holds the hub and the shaft together is quite literally the forces of nature (aided by some significant hydraulic pressure during installation). This hub had failed (spun on the shaft) three times on startup. Each failure resulted in a one year repair to the pump as it had to be shipped from the desert to Belgium. The cost of each incident was $1,000,000. There was a real interest in solving this issue.

Zeroing In on the Root Cause

The Cause-and-Effect diagram is shown in Figure 3-7 on the following page.

One of the side benefits to constructing a cause-and-effect diagram is that you actually are assembling a troubleshooting decision tree. This guide can essentially be used to 'run out' potential causes for several abnormalities throughout the life of a machine. Also, this product can continue to grow in possibilities as more and more issues are discovered and addressed. It can be a very valuable resource.

Spare the Rod

A final thought on the subject of Root Cause Analysis. During my military service, I served additional duty as an aircraft crash investigator. Of the many incidents I investigated as the maintenance member, I was always pleased to know that the facts and conclusions the board drew could not be used in a punitive manner against anyone. If the Air Force wanted to hold someone responsible for a crash or incident, they (the Air Force) had to hold a completely separate investigation. What our accident investigation boards were commanded, by law, to do was to determine the cause(s) to keep the situation from ever repeating. The military learned a long time ago that people speak more honestly when they know they are not going to get punished for the truth.

RCA, like problem solving, is a helpful and greatly beneficial practice. Ask yourself if the members of your organization look at RCA and problem solving that way and if they eagerly look for opportunities to be on an investigative board.

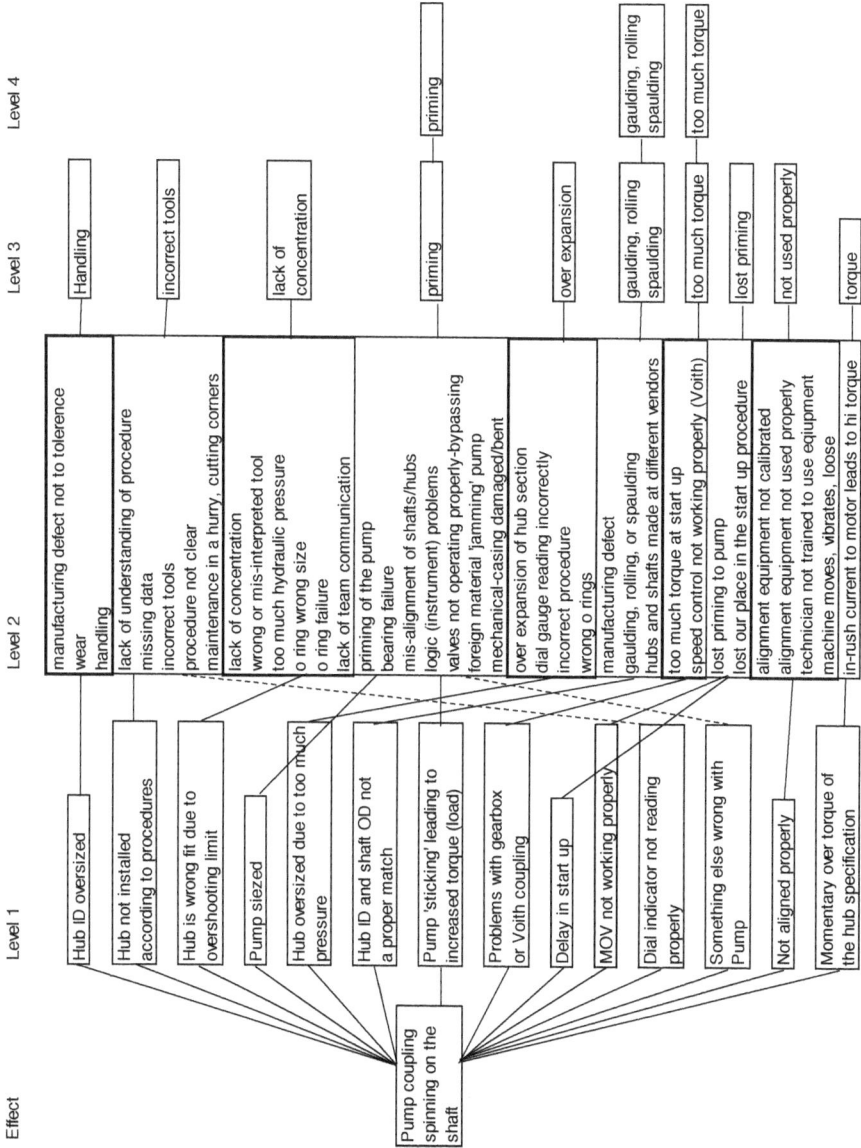

Figure 3-7:
Cause-and-Effect
diagram for
hydraulic
coupling

Define the Problem Correctly and the Solution Will Be Staring You in the Face

Problem solving is as much art as it is science. Albert Einstein is said to have advised that if you have an hour to solve a problem, he'd recommend spending fifty minutes defining it. If you've ever heard the phrase before taking a test, "Read each question carefully," you know what Einstein was hinting at. Often the answer is within the question itself.

Here are some examples of questions I've been asked to consult on utilizing a formal problem-solving approach. Following each example is my smart-aleck response to the client. I actually said these things:

> Client (C): "The storeroom never has anything."
> Me (M): "Really, the storeroom doesn't have anything?"
> (C): "That pump fails all the time."
> (M): "The pump never works?"
> (C): "The operators break the equipment all the time."
> (M): "You mean the plant is down right now?"

Clearly, these are extreme examples, but they are more common than you would think. In fact, you may have heard some of these yourself. Our trouble isn't that we can't solve problems, it's usually that we can't agree on what the problem really is. If we could get really good at defining problems, we'd be awesome at solving them.

A colleague told me one time that world-class organizations are world-class problem solvers. I believe him. Defining a problem is the key to solving it, so let's become exceptional at defining a problem. The critical questions to ask during the definition phase are:

1. Who discovered the issue (who)?

2. How big is the issue (what)?

3. Where was the issue (where)?

4. What time frame was information collected in (when)?

5. The issue resulted in what, exactly (resulting in)?

A Common Scenario as an Example

Here is an example of parts being taken from the storeroom without the technician documenting the withdrawal:

The day shift clerk recorded eight incidents where items were removed from stores without recording the work order number or the person that removed the items. This was recorded over a two week period, resulting in the absence of issuing documentation.

- Who discovered the issue? —**The day shift clerk**

- How big is the issue? —**Eight incidents in which items were removed from stores without recording the work order number or the person that removed the items**

- Where is the issue? —**The storeroom**

- What time frame was information collected in?—**Two weeks**

- The issue resulted in what, exactly? —**The absence of issuing documentation**

The result of the problem is significant for the sole purpose that it tells us how much we care about the problem. Someone could determine that this particular consequence has no real effect, and the problem is not worth solving.

However, if the consequence is looked at from the perspective that, because of the absent issuing documentation, several items never 'hit' their reorder point, it can be seen that the ultimate result is that the storeroom never has anything.

Now it's time to polish your skills at problem definitions, or in other words, writing a problem statement.

Exercise

The scenario:

The support team lead technician has the primary responsibility of ensuring that the plant compressed air is held at eighty PSI during all phases of plant production throughout the facility. The support team is responsible for the plant's two air compressors. During normal operations, one air compressor is running all-out, and one is trimming over the top. The manufacturing facility operates 24/5, choosing not to run during the weekends and holidays. The support team ensures the air compressors are on-line two hours before startup on Monday mornings. Operations conducts a staggered startup protocol. The lead has noted that on three separate occasions, the two plant air compressors have failed to maintain a plant air pressure of at least eighty PSI. These incidents have occurred over the last four weeks, and each time was during plant startup on a Monday morning. Because of this issue, there has been a repeated delayed start of 20% of the plant's machinery, resulting in at $20,000 loss of potential revenue.

Answer these five questions:

1. Who discovered the issue (who)? _____

2. How big is the issue (what)? _____

3. Where was the issue (where)? _____

4. What time frame was information collected in (when)? _____

5. The issue resulted in what, exactly (resulting in)? _____

Using the answers to those five questions, develop a practical problem statement:

_____

_____

_____

Here is what I came up with:

The support team lead technician has noted that on three separate occasions, the two plant air compressors have failed to maintain a plant air pressure of at least eighty PSI. These incidents have occurred over the last four weeks, and each time was during plant startup on a Monday morning. The result of this issue has been the delayed start of 20% of the plant's machinery, resulting in a $20,000 loss of potential revenue.

Never again write a problem statement as, "The air compressors never hold plant air pressure." You now know better!

A closing note on RCA and problem solving. If at all possible, do not conduct these investigations unilaterally. Always assemble a team of engaged associates. Sometimes the value in solving a situation isn't the result, it's the experience the team gets at working together to solve something significant.

KPIs Aligned With Goals

KPIs and goals were addressed in great detail in Chapter 2. We are going to discover in this section the fundamental error most organizations make in aligning the two and how we can correct this matter quickly.

First, recall that we established a very critical connection linking goals to means to consequences. The *consequences* were the target metric we hope to achieve. The entire effort is realized by the execution of the *means* with which we achieve the desired consequences.

That's enough of a memory jogger. Let's consider some *poorly* aligned goals and KPIs.

Goal: Increase production output
Metric: Run longer (KPI—run time)

Goal: Improve safety
Metric: Hold more mandatory safety meetings
          (KPI—mandatory attendance at meetings)

Goal: Improve safety
Metric: Conduct more safety inspections (KPI—number of
          safety work orders generated)

Goal: Improve quantity of PMs conducted by demanding
          100% compliance
Metric: PM count (KPI—PM Completion)

Goal: Increase uptime
Metric: Less downtime (KPI—less maintenance)

What do each of these poorly aligned goals and KPIs have in common? The plain truth is the activity being measured will not deliver the desired goal.

Why? Because there is no leading indicator telling us if we are doing the activities that drive the lagging indicators. The leading indicators are often the 'means' which we have discussed previously in the section on Goals.

Before maintenance can align their goals with their KPIs, there must first be an alignment of the reliability goals with the plant, facility, or company goals.

Check Your Alignment

Here is an example:

Company goal: Decrease *Cost Of Goods Sold* (COGS) by 10% this fiscal year

Think about that for just a minute. Without any mention of capital investment, the company is requesting that the plant decrease COGS by 10%. For the team that should indicate that the equipment needs to run more efficiently, more effectively, and perhaps run more often than not.

In order to achieve this, the maintenance department needs to take the machinery down less often, and do better maintenance work in shorter time. This requires maintenance work that is planned, scheduled, kitted, and executed right the first time— maintenance that sticks. It also will require that the storeroom store the spare parts in a manner that protects the inherent reliability of the components.

In the past, we might have had a goal and KPI alignment that looked something like this:

Decrease COGS by 10% by cutting maintenance and spare parts.

By establishing the link of goals to means to consequences developed earlier, we now know that our alignment should look something like Figure 3-8 on the following page.

This model of goal and KPI alignment can sometimes be shown as a *Balanced Score Card* (BSC).

In our example, the BSC can establish a direct line between reliability efforts and success for the company. It is in many

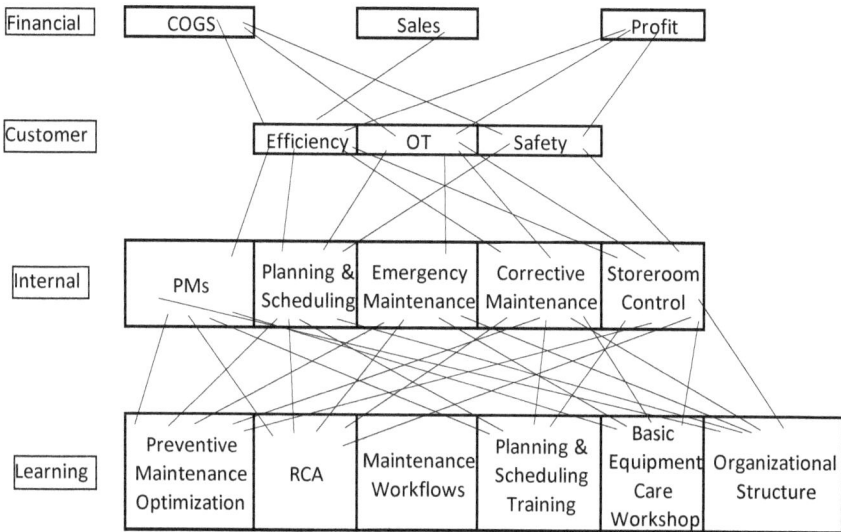

Figure 3-8: Balanced Score Card (BCS)

ways the missing link. The corporate vision and strategic interest are required to make this connection; these are very important. "It cannot be emphasized enough that it is fundamental that the balanced scorecard be based on the comprehensive vision and overall strategic aims of the organization (N. Olve and A. Sjöstrand, *Balanced Scorecard*, p. 100).

"A rising tide lifts all boats"—John F. Kennedy

A BSC takes into consideration the customer's objectives first, and then builds from there. I never refer to production as the customer except in this example. Maintenance must focus on what production needs and wants. After all, maintenance exists to provide exceptional technical advice and expert wrench work to production.

After the customer's needs are annotated, then the reliability group makes note of what kinds of activities are needed to deliver what production has stated (in writing) they need in order to comply with their directives. These are the internal objectives.

Once production and maintenance have their requirements listed, then the financial impact needs to be documented. The financial impact takes the form of what the company demands.

To complete the Balanced Score Card, there is one more link, and that is a learning area where maintenance lists the types of training and capability it must have in order to be able to deliver on what it has promised.

The rule with a BSC is that each individual item must be linked to at least one item in the level above and the level below it. This is a direct alignment of goals to KPIs.

Figure 3-8 is an example from an actual reliability improvement strategy laid out as a Balanced Score Card. From this example, it is clear to see that operations desires to have greater efficiency. In order for that to occur, maintenance must conduct high performing PMs. However, in order for maintenance to ensure their PMs are of the highest quality and effectiveness, maintenance must perform Preventive Maintenance Optimization (PMO) work on their existing PMs. The level indicating the PMO works is the *learning* level.

Look at that score card again. Can you see the link between the *maintenance-internal* objective of conducting PMs and the *production-customer* objective of having greater efficiency? In this example, if production doesn't provide access to the equipment for scheduled preventive maintenance, then there is no guarantee of availability. Production shouldn't expect top running equipment if they don't surrender the asset from time to time. This is absolutely the definition of aligning goals with KPIs.

The KPIs become the measure of our ability to execute the goals listed on the BSC. For PMs, our KPIs could be:

- PM compliance

- CM/PM (Corrective Maintenance from Preventive Maintenance)

- PMO workshops completed

- Assets which had a PMO workshop conducted vs. total plant assets

- MTBF

These KPIs are a mixture of leading and lagging indicators.

What a Real BSC Might Look Like

By the way, just as a demonstration of the clean examples often provided in books vs. the actual field work, take a look at Figure 3-9. This is the same BSC, in its raw form, as it appeared after the strategy session at the customer's manufacturing plant.

Figure 3-10, shown on the following page, provides some additional information and rounds out the BSC. Once we have the connections of goals and objectives shown in Figure 3-9, we still need to lay out the measurements that we are going to use as the KPIs, the target (consequences) and the initiatives (means) to take to make all this possible. This is the goals to means to consequences connection that we've been studying. Can you find the means and consequences sketched out in Figure 3-10?

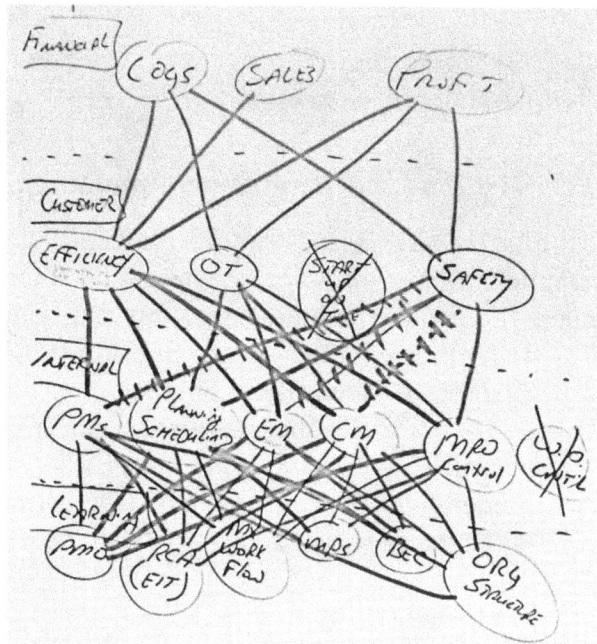

Figure 3-9:
The real BSC,
in the field

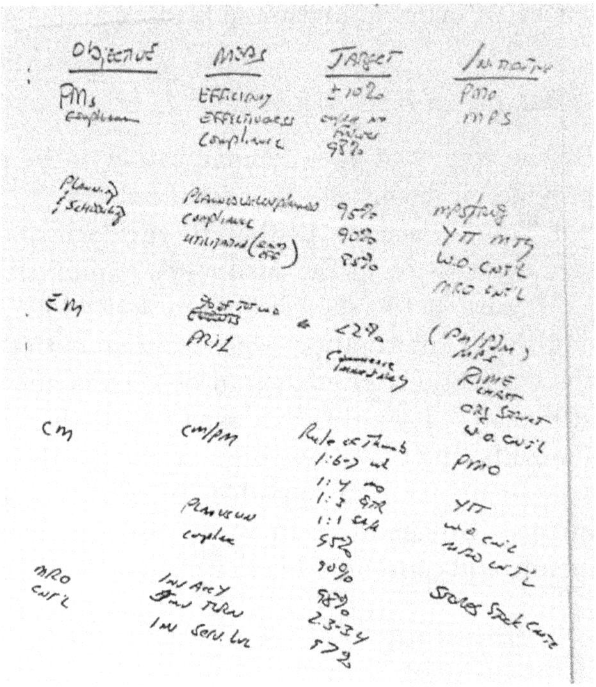

Figure 3-10:
Additional BSC
information

Yes, you've just seen my actual handwriting. I like to think that passion guides my pen.

Let's Try Your Pen at an Exercise on Balanced Score Cards

Figure 3-11 is a Balanced Score Card I started for you to complete as an exercise. I've listed what the customer (production) needs, and what the financial (company) requirements are as well. For this exercise, make entries in the internal (maintenance) and the learning sections based on what you believe would be effective processes and policies. Next, make a direct line connection between all the layers and processes, as demonstrated in Figure 3-8. Remember, each process must be connected to at least one process or policy in the layers above and below.

Finally, take one singular activity from what you've entered in the internal (maintenance) level and list at least two KPIs that can be used to measure the execution success of that activity.

Finally, take one singular activity from the internal (maintenance) level and list at least two KPIs that can be used to determine the degree of execution for that activity (like in Figure 3-10).

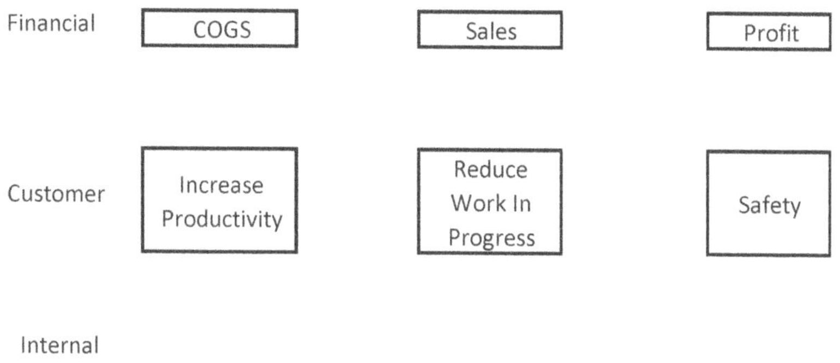

Financial	COGS		Sales		Profit

Customer	Increase Productivity		Reduce Work In Progress		Safety

Internal

Figure 3-11: "Exercise for BSC"

KPI #1: _____

KPI #2: _____

Once you have completed this exercise, you have effectively aligned the goals of the company with the goals of both production and maintenance, and established the KPIs that tell the story of how well the organization is doing at attaining their goals.

Good job!

MOC

In the last section of this chapter we are going to discuss PSM, or *Process Safety Management*. One of the fourteen elements of PSM is *Management Of Change* (MOC). A Management Of Change process is perhaps one of the easiest and most agreeable programs to understand and expect, but somehow we've managed to mess up the execution of this very important program and we've negated much of the benefit that might have been gained.

An MOC actually provides two functions. Primarily, the MOC is a coordinated, formally documented process that is used to determine if a 'change' is technically feasible. For example, if a manufacturing plant was considering changing a crucial valve from hand-operated to a control valve, an MOC would be used for all interested parties to determine if such a change had an effect on production, quality, safety, performance, or the process as a whole.

An MOC is also used to inform all the concerned parties that a change is taking place, so it becomes a means to communicate.

Have you ever returned to work after a vacation and something had changed? Were you surprised when you discovered that there was a change and no one told you about it? An MOC eliminates that surprise.

Where the MOC process becomes complex is in the bureaucratic administration of the process. It becomes a red-tape jungle. Without serious dedication and discipline to the process, it would not be uncommon for a change to be held up because the MOC administrator is on vacation.

Without deliberate attention to the MOC program, an entire project could be delayed up to the point where the paperwork is pencil-whipped just to get the project back on track. This could be catastrophic.

As we'll discover later in this chapter, Process Safety Management is focused on processes that utilize or affect one of the 155 *Highly Hazardous Chemicals* (HHCs) identified by OSHA. MOC, as stated, is one of the elements of PSM. I would strongly recommend that even organizations that don't have any of the HHCs identified by OSHA create and monitor an MOC program. Again, consider that the MOC is used to determine if a change is technically feasible and to communicate the change that is taking place.

As an example of a practical use of the MOC process, list one change that has recently been made at your location. This change could be a physical component change (like a valve style change), or a process change (like reducing cooking time, or holding time). Note that a like-for-like component change does not constitute a *change* as it relates to an MOC.

List your change here: _____

In the space provided, list the agencies that would be concerned with the change that you've listed. The MOC would be routed to these individuals for the change that is being addressed in the example.

_____

_____

_____

If you currently have an MOC process, act now to enhance it. If you don't have an MOC process, get one and administer it like your safety depended on it. You'd be surprised how real this warning might be.

Regulations and Rules

Any discussion in a maintenance book about regulations and rules is warranted as it is often these mandates that dictate what is written into the preventive maintenance program.

Deming has a lot to say about such specific directives, lending his pulpit and voice to flatly tell us that it's through the structured and systematic adherence to something exact that helps control our processes. "Once a process has been brought into a state of statistical control, it has a definable capability. It will show sustained satisfactory performance..." (*Out of the Crisis*, p. 339)

It is, Deming continues, the catalyst to a bevy of positives, "A process that is stable, in statistical control, presents a number of advantages over instability. In statistical control:

- The process has an identity; its performance is predictable...

- Costs are predictable...

- Regularity of output is an important by-product of statistical control

- Productivity is at a maximum (costs at a minimum)" (*Out of the Crisis*, p. 339)

To a great extent, rules and regulations help to establish a degree of statistical control over our reliability processes.

Example

For example, give some thought to the *safety relief valves* (SRVs) on a boiler system. OSHA and state safety agencies indicate that at least an annual inspection of the SRVs will be conducted by manually lifting (unseating) the valve and releasing it. The valve should reseat upon release. If it doesn't release, you can try one more time. If the valve does not successfully reseat on the second try, the valve must be replaced. These valves can often be sent off for rebuild.

This is a safety regulation, and most likely a rule in your organization. Often, a rule complements a regulation as insurers are brought into the discussion.

Almost all organizations that perform a maintenance function have some form of *Computerized Maintenance Management System* (CMMS). Whether you know it or not, the CMMS documentation of the safety and performance checks and maintenance of your OSHA and EPA regarded equipment is THE only documentation OSHA or the EPA is likely to accept as proof that you are maintaining the equipment properly. Again, regulations and rules.

Determine Credibility

This next point is meant to be confrontational and debatable. I think I'm right, though. When a rule or regulation is handed down to maintenance or the reliability group, it is fair and expected that the person in charge of equipment reliability should request, "Show me that in writing."

The reason the maintenance manager, or person with overall responsibility for reliability should ask this question is to align what is required with what is to be executed.

Imagine if you were not aware of the exact wording and reference of the SRV comments made earlier. How would you even know if you were complying with what was required? How would you be certain that you weren't doing more than that which was required (inefficient)?

In the military, we got really good at starting orders and other directives with 'IAW,' meaning In Accordance With.

It is very useful to cite the regulation or rule in each and every maintenance task, when applicable. Also, these maintenance tasks need to be tied administratively to any change in the regulation or rule via an MOC.

Where Do These Directives Come From?

Regulations and rules can come from many sources. Just to name a few:

- The federal government

- State government

- Local and municipal government

- Parent company (HQ)

- Plant or facility leadership

- Engineering

- Safety

Another source for rules and regulations comes from OSHA's Process Safety Management in the form of an acronym, RAGAGEP, meaning *Recognized And Generally Accepted Good Engineering Practices.*

Have you ever wondered what the source was for the practice of only torqueing a bolt one time? How about the fact that we will put a new belt on an old sheave, but we won't put an old belt on a new sheave. RAGAGEP.

Almost all trade practices and common sense elements of maintenance execution can be traced back to a recognized, vetted, and respected source. These sources might include:

- *National Fire Protection Association* (NFPA)

- *National Electrical Code* (NEC)

- *International Plumbing Code* (IPC)

- *American Society of Mechanical Engineers* (ASME)

All these sources should be referenced in our maintenance tasks. "Reference in a regulation to an industrial standard provides a line that makes the regulation effective and meaningful." (W.E. Deming, *Out of the Crisis*, p. 304).

Failure to list the defining reason something is required and to what degree it should be executed invites the technicians to take the work less than seriously. This could lead to dramatically poor reliability, not to mention safety.

Are Your Tasks Linked to Anything?

In the space provided, list one singular PM task. Next to that task, list the source requiring that task. It could be as simple as a recommendation from the OEM.

Task: _____

Source: _____

We have got to get into the practice of confirming the need for maintenance tasks based on a respected and documented regulation, rule, or trade practice. Maintenance is most efficient when we aren't just doing something because "Jerry said to do it."

PSM

Process Safety Management (PSM) has been discussed many times in the previous sections, so the build-up and anticipation should be ripe.

A Federal Regulation

PSM is a directive established by OSHA in 29 CFR 1910.119. The section in the *Code of Federal Regulations* (CFR) is aptly named *Process Safety Management*. The intended use of this CFR is to provide direction and process safeguards to companies that use one or more of the 155 identified Highly Hazardous Chemicals (HHCs). The regulation lists the HHCs that fall under this jurisdiction, and establishes the threshold levels that make a particular chemical controllable.

Food industries often find themselves under the auspices of PSM due to their use of anhydrous ammonia. A level beyond 10,000 pounds would make the chemical identifiable under PSM and a formal process is required. In fact, if a company has more than one HHC at or above the listed threshold level, a separate PSM program is necessary for each individual chemical.

Of course this is a federally regulated requirement and, as such, is under the scrutiny of OSHA and could be audited at any time. A word of caution here. This is one of the few areas where

OSHA and the EPA work together. For instance, if the EPA is auditing your location, they could observe a situation regarding the HHCs that would allow them to bring in OSHA. Also, if OSHA is at your facility for any reason at all, they could ask to see your documentation for your PSM program(s). This is a very complex area of administration. Read that as meaning 'lots of paperwork.'

We're Here to Help

Years ago, I made an appointment with the OSHA office in the town where I lived at the time. I actually was able to speak with the agent in charge of this particular office. I told her that I was seeking some information on PSM to write an article in one of the maintenance trade magazines, and to also put together a lecture and course presentation on the subject. This is a direct quote from the agent in charge: "Why would you want to do that?"

I want to make this unmistakably clear: we must have a solid PSM program to ensure the safety of our people, our facility, the community at large, our brands, the environment, and our company. This is not a battle cry against these federal agencies. Rather, this is a rallying cry to all reliability professionals and our production and leadership organizations to say, "We need a rock solid program, well planned and well executed." This is the real deal.

The Regulation is Very Informative

There is much information in 29 CFR 1910.119. In fact, it is perhaps one of the best written, and unambiguous government documents I've ever read. This section isn't designed to make you an expert on the regulation, but it is an introduction to PSM. Listed below are the fourteen elements that make up the regulation:

1. Employee participation

2. Process Safety Information (PSI)

3. Process Hazard Analysis (PHA)

4. Operating procedures

5. Training

6. Contractors

7. Pre-startup safety review

8. Mechanical Integrity (MI)

9. Hot work permit

10. Management Of Change (MOC)

11. Incident investigation

12. Emergency planning and response

13. Compliance audits

14. Trade secrets

To conclude this section and Chapter 3, review Table 3-4. This table shows how OSHA has divided the fifty states and U.S. territories into ten manageable regions.

I'd strongly recommend that companies and locations that are not under the auspices of OSHA's PSM program consider forming a PSM process anyway. Process Safety Management provides a tremendous source of leading activities that ultimately results in a safer and more engaged workforce.

Review Table 3-4 one more time. See if you can fill in the following spaces with the proper information:

Table 3-4: OSHA's ten regions

Region 1	Region 2	Region 3	Region 4	Region 5	Region 6	Region 7	Region 8	Region 9	Region 10
ME	NJ	PA	KY	MN	NM	NE	MT	CA	AK
VT	NY	WV	TN	WI	TX	IA	ND	NV	WA
NH	Puerto Rico	VA	NC	IL	OK	KS	SD	AZ	OR
MA	U.S. Virgin Islands	DE	SC	IN	AR	MO	WY	HI	ID
RI		MD	GA	MI	LA		UT		
CT		DC	AL	OH			CO		
			MS						
			FL						

What region are you in?

_____

What is the address and phone number of the OSHA office in your area?

_____

Who is an agent in that office that can be a point of contact for you?

_____

Can you request a copy of 29 CFR 1910.119?

_____

Chapter Summary

This chapter on the Process of Reliability is an interesting study, if for no other reason than it concentrates on the need to write stuff down. I would concede that an organization which has no formal and documented strategy and processes can have a good maintenance department. I can agree with that, if you would agree that in order to have an excellent maintenance department, there must be some formality with processes and policies.

Here at the end of Chapter 3, let's remind ourselves that we are building a compelling strategy and outline for a world-class maintenance organization. We covered Change and the Resistance to it in Chapter 1, and the Business of Reliability in Chapter 2.

Now we are going to continue to build on the previous work toward our strategy with Chapter 3's contribution.

In Chapter 3, we discussed:

- Ownership
- Commitment vs. Compliance
- Workflows
- RACI Chart
- Process Guides

- Roles and Responsibilities

- Continuous Improvement

- Problem Solving-RCA

- KPIs aligned with goals

- Management of Change

- Regulations and Rules

- Process Safety Management

Our developing philosophy:

The maintenance and reliability organization will strive to establish a sense of process ownership within our associates. We will do this by marrying responsibility with authority. Our desire is for commitment over compliance, believing that ownership is the catalyst for commitment.

The processes that we will own together with our partners are:

- Identify
- Approve
- Prioritize
- Plan
- Schedule

- Turnover
- Execute
- Return to service
- Closeout
- Evaluate

The storeroom processes that we will own are:

- ABC Classification Review

- Adjust ABC Classification Model

- Add to Stock

- Bench Stock

- Critical Spare Parts Algorithm

- Critical Spares Review

- Cycle Counting Criteria

- Cycle Counting
- Data Scrubbing
- Disposal of Scrap
- Document Stockout
- Emergency Procurement
- Issuing
- Item Substitution
- Kitting
- Obsolescence Criteria
- Obsolescence
- Optimizing Stock Levels
- Ordering
- Preservation Program
- Receiving
- Repair or Replace
- Return to Stock
- Return to Supplier
- Salvage Value
- Special Tools
- Standardization
- VMI/Consignment

Each of these processes will be formally documented in process guides. All process guides will include the following sections:

- Business and Process Description
- Applicability

Business Process Overview

- ⌐ Inputs
- ⌐ Outputs
- ⌐ Touchpoints
- ⌐ Key design decisions, requirements, and expectations
- ▪ Business Process Workflow
- ▪ RACI chart
- ▪ Activity detail

Formally documenting these reliability processes will further enhance our ability to properly staff the efforts of the maintenance department by aligning the actual needs of the process with the roles and responsibilities of the associate.

From the processes, we will garner a sense of clearly defined expectations, leading to the creation of Key Performance Indicators that are in line with corporate, plant, and reliability goals. The KPI and goal alignment will be created using the balanced score card method, a sample of which is shown.

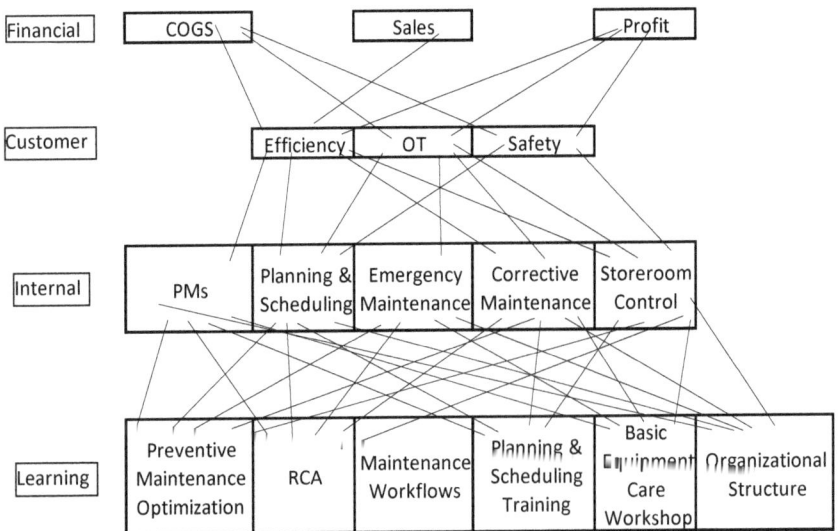

The net gain from this squaring of human resources to formally established processes, as well as expectations to KPIs, will be a maintenance organization that is a value added contributor to overall culture change.

We will aggressively attack problems by developing an exceptional problem statement technique. Our problem solving statements will take the form of:

- Who discovered the issue (who)?

- How big is the issue (what)?

- Where was the issue (where)?

- What time frame was information collected in (when)?

- The issue resulted in what, exactly (resulting in)?

Our maintenance organization is committed to the weeding out of issues, unfavorable trending of key metrics, and equipment reliability deficits. As such, we will employ the Root Cause Analysis (RCA) tool(s) of:

_____

_____

The processes that we will adhere to in order to maintain good order and discipline to high performing maintenance will be controlled through the process of Management Of Change (MOC). As part of the MOC program, any change that is not like-for-like will be vetted for technical feasibility and will be well communicated. This includes any change to rules and regulations.

Where possible and when applicable, we will research and cite rules, regulations, and suggestions for maintenance tasks to demonstrate the actual need, along with the actual process. This is meant to contribute to the ownership practice we will enhance in maintenance. As previously stated, these sources will be managed and controlled.

We further pledge to enhance, or create, a Process Safety Management (PSM) program to guarantee employee engagement and control of all safety and process requirements. This will be a leading indicator of our personal safety program.

That concludes Chapter 3. You're halfway through this book, and you've developed half of an exceptionally pragmatic maintenance strategy. Well done!

FOUR

Ensuring Reliability

"On the Nautilus, men's hearts never fail them. No defects to be afraid of,
for the double shell is as firm as iron; no rigging to attend to;
no sails for the wind to carry away; no boilers to burst; no fire to fear,
for the vessel is made of iron, not wood; no coal to run short, for electricity is the
only mechanical agent; no collision to fear, for it alone swims in deep water; no
tempest to brave, for when it dives below the water, it reaches absolute
tranquility. There, sir! that [sic] is the perfect of vessels!"

—Jules Verne, *Twenty Thousand Leagues Under the Sea (p.71)*

In Chapter 4 we are going to explore some of the broader aspects of maintenance and reliability by discussing some elements to an effective approach that are often 'swept under the rug.' Quite honestly, these particular discussion points are not taught anywhere in a central learning location. They are most certainly not taught to the general maintenance population in most organizations.

A central theme to this section is that you basically get out of something what you put into it. Equipment seemed to be more robust in the past, and certainly the electronics and technology world was not as fluid as it is today.

When Captain Nemo was commanding the world's oceans as a pseudo-aquatic Robin Hood, he did so with a marvelous vessel. This ship was way ahead of its time in terms of technology (the book was written in 1870), but it was more than that. It was a ship that was designed and built for the intended purpose. Imagine building a machine to operate as intended, and to survive in the grueling environment in which it operates.

Although a work of fiction, Verne's work was not far off from what is needed today. This might be a stretch in logic, but his

(Nemo's) was a machine that was built free of defects, essentially problem free. It was built for the environment and the purpose in which it was intend to be used. It was indeed "the perfect of vessels!"

For equipment to be built to last, it has to first be designed to last.

Equipment Design Excellence

Design for excellence might be a better term to focus in on when it comes to what is truly important for giving maintenance and reliability professionals a fighting chance to be successful. If the foundational principles of building an asset or the modification to an asset are done correctly at the start, good things can happen.

I'm fond of saying that if you design a piece of equipment poorly, build it cheaply, and install it hastily, you aren't going to have a world-class asset on your hands. You are going to have a nightmare for the next forty years!

A similar sentiment is shared in *TPM in Process Industries*:

> "Particularly in process industries, major equipment items are often customized to individual specifications; they are often designed, fabricated, and installed in a rush. Without strict early management, such equipment enters the test operation phase with many hidden defects. The truth of this is borne out by the frequency with which maintenance and production personnel discover defects generated in design, fabrication, and installation during shutdown maintenance and startup." (T. Suzuki, p. 199)

In the introductory section of this workbook, you were asked to provide your interpretation of a few common terms. I'd like you to rewrite two of those here:

Reliability:

_____

_____

Maintainability: .

_____

_____

These are pivotal terms that come into the conversation when we are talking about equipment design excellence. Equipment arrives at our location with inherent reliability. You can't maintain more reliability into an asset than it has inherently.

The mission of maintenance then becomes to maintain the inherent reliability of an asset. We ensure the reliability by enhancing the maintainability.

$D \times B \times I$

I introduced a quip in an earlier paragraph: if you design a piece of equipment poorly, build it cheaply, and install it hastily, you aren't going to have a world class asset on your hands. You are going to have a nightmare for the next forty years. Let's show that algebraically:

$$D \times B \times I$$

Figure 4-1: Design x Build x Install

We need to show this algebraically to indicate that an improvement in any of the three integers (D, B, or I) will improve the overall reliability, whereas the slightest miscue in any could cause a problem with the whole.

This formula isn't calculating inherent reliability of an asset, but what it does do is calculate our chance of having high reliability. Let's further our discussion by equating this, somewhat loosely, with calculating a system using block diagrams.

Using the Formula
For an example, let's assume the following values:

For *D—design*; we've got almost a perfect, flawless design. This is a rock solid product. We'll give it a 99%.

For *B—build*; we've got the best OEM for this product, this is a five star manufacturer. Just to game our system a bit, let's give that a 99% as well.

For *I—install*; we are using our trusty ol' general contractor who has been in this plant for years. The old man himself is coming out to supervise this install. Everything is perfect so it also gets a 99%.

Using the block diagram approach for calculating our reliability chances, we have:

$$(.99 \times .99 \times .99) \times 100 = 97\%$$

We've done our part to design, build, and install the best machine possible. It's near perfect, actually. Now it is up to operations and plant maintenance to keep the asset humming at this exceptional level.

But wait a minute–things never go as planned. Even the best intentions have to yield to the project budget.

We likely set out to design the best machine that is possible, not initially considering the cost. After all, let's at least ask for what we want. Our design, if we are doing it correctly, would hope to utilize the parts we already have in our storeroom. These are parts that, for whatever reason, seem to survive in our environment. Why wouldn't we want to build those into our new asset? This is standardization.

Our initial design might call for top-shelf components. We would plan for redundancy and built-in diagnostics. It's likely that our design would even include 110-V convenience outlets. Ours would be a fine machine.

Now comes concession time. We have to carve and round out some features to fit into the budget. Even the capital spares idea goes out the window. Pretty soon, our *D—design* goes from a 99% to an 85%.

The *B—build* doesn't fare much better. We're forced to go to the lowest bidder and this alone knocks us down a few percentage points. As we discover during construction, the component we actually wanted doesn't physically fit, so we have to get another model that is almost as good. Since we didn't specify the exact components we would accept, we end up with the components the OEM gets the best deal on, and 99% becomes 82%, just like that.

I—install is the end of this adventure. We are utilizing space cleared from another machine, and we do our best to clean up the site. Someone has the cost saving idea to reuse the old disconnect and power distribution panel, saving at least $1,000 against our budget.

Earlier we had the bright idea to prepare the site with a vibration isolation pad, but we ran out of money. And the cooling tower is a little under-sized, but it's bound to work.

We had two weeks to install this unit, but the plant manager decided to run six days into that. Now we have eight days (one of those is Thanksgiving).

For all our trouble, *I—install* falls from 99% to 87%.

I refer to these values as the EDE rating, or *Equipment Design Excellence rating.*

Our calculation now stands at: .85 x .82 x .87 = 60.6% (a 60% drop)

Where did it all go wrong?

Much of this can be improved, and be guarded against any decrease, by having set processes and procedures for capital projects and major repair projects, and by sticking to those processes.

Also, we can give ourselves a better chance of success by having set component standards that we insist on. This helps not only the equipment design excellence effort, but the storeroom and the maintenance group in delivering maintainability.

There is a significant difference in what it will cost and how many joules of human energy will be needed to maintain an asset that comes at us with a 97% EDE rating as opposed to a 60.6% EDE. That difference is 60%!

A note on where the numbers come from. How do you determine if the design is an 85% or a 90%? This is a fairly open scoring system, but one that does need to have some rules around it.

In fact, as part of the capital and major repair process development, I'd suggest that the group engaged in developing the processes further develop an unambiguous scoring system that can be used on all projects. The scale will not change. It is carved in stone. This is a system that is particular to you, your organization, and your company.

Let's run a very quick exercise to demonstrate the power of this number.

Give This Exercise a Try

Write down the last capital project or major modification conducted in your facility: _____

On a scale from 0% to 100%, give a value to each of the three following categories:

Design: _____
Build: _____
Install: _____
Overall D x B x I = _____

Chances are, if the DBI number is <60%, the asset has been hard to maintain and troublesome. If the number is >80%, it has been operating fairly well, and is not much of a burden. A number that is in between 60% and 80% is common territory and, honestly, the value we are likely to calculate for the equipment in our facilities. These 'in between' assets will run fairly well by our better operators. Lesser skilled operators, or those with poor attitudes will seem to have more issues with these machines.

If you were conducting a post project review of the project you just evaluated, what suggestions would you make to arrive at a higher D x B x I value?

_____

_____

_____

_____

_____

How Is a New Asset or Modification Created?

This might be the most obvious point in this workbook, but the best design for an asset or a machine modification comes from those people who are the most affected by the new creation. Clearly, there needs to be parameters established for the cost, performance, size, and general layout of the equipment, but the associates working in and around the machine need to have a seat at the table when there is but an inkling of an idea that something new is intendod.

Ron Moore said as much in *Making Common Sense Common Practice*:

> "A focus group consisting of maintenance personnel (mechanics and engineers), operations (operators and production supervisors), purchasing and stores personnel, and design/capital project engineers should be convened. After an introduction as to the purpose of the focus group—getting operations and maintenance input for the coming design—these case histories should then be presented to the focus group to start the group thinking about where the potential opportunities are and what might be done to capture them. The goal is to identify major opportunities for improving plant reliability through "designing out" current problems in existing operating plants and thereby reducing life-cycle cost (as opposed to installed cost). As a bonus, you also get higher uptime, lower unit cost of production, and the opportunity for higher market share and operating profits." (p. 120)

Inherent Reliability

We introduced the term *inherent reliability* in the preceding section. We also discussed a theory that you can't maintain more reliability into a machine than it has inherently. In fact, as we stated, the role of maintenance is to guard the inherent reliability of the asset. One of the storeroom's responsibilities, we'll find out later, is to guard the inherent reliability of the components in the storeroom.

What is inherent reliability and how do you get more of it?

Defining Inherent Reliability

Consider that every machine you have in your plant or facility is made up of the same components. Every asset has some of the following:

- Valve
- Coupling
- Sprocket
- Chain
- Sheave
- Belt

- Gearbox
- Cylinder
- Button
- Controller
- Switches
- Wiring

- Fasteners
- Frame
- Handles, knobs, levers
- Zerk fitting

Choose one asset (equipment) in your facility and write it down here: _____

Now, go back to the list of components just listed and highlight those components which appear on the asset you just recorded.

In Table 4-1, record those component types in the column titled Component Type.

We're going to stay on Table 4-1 to add some more information. In the column titled "Current Brand, Make, Model," indicate what is currently on the machine in regard to each component type. If you don't know this information, that is another issue all together.

Table 4-1: Asset components

Asset:	Component Type (e.g. motor, cylinder)	Current Brand, Make, Model	The Absolute Best Brand, Make, Model

In the final column, see if you and a small team can agree on what would be the best brand, make and model for each of those components. Give some thought as to what you would buy if money was no object.

The resultant table draws the contrast between what you have as far as a level of inherent reliability vs. what you could have. In fact, this ties nicely back with the *D—design* and *B—build* discussions in our previous section.

It All Adds Up

Inherent reliability is the cumulative reliability of each component that makes up the asset. We have to take into account that there are certain components that, for whatever reason, seem to work in our crazy environment. There might not be any rhyme or reason. These particular items just seem to be robust enough to survive.

This brings up a very interesting point regarding reliability. A lesser component might actually fare better in the situation that we expose it to as opposed to a more highly regarded component.

Almost universally, folks can agree that they've seen an old cheap part outlast some of the new, higher tech models.

When we consider what components would be best to use on our new asset, we have to give some regard to what seems to work really well 'around here.' This touches on the idea of standardization. As much as possible, let's stick with the components that survive longer in our facility, and demand that we standardize around those.

It bears repeating that an OEM will construct your next machine with the components they get the best deal on. OEMs don't actually make components. They make the frame and put it all together. Very few OEMs make bearings and motors, etc. We tell them what they are allowed to put on our machine. We control the inherent reliability.

Cost is just one of the overall concerns when considering a new asset or the modification to the old one. Our decisions today will impact our employees forty years into the future. Not just with the frustration they are likely to have with the equipment we are installing now, but with the cost those in the future will have to bear in maintaining the assets. Cost can affect us now, and our company well into the future.

Recall that Gulati calculated "...O&M cost is on average about 80% of the total life cycle cost of the asset. It is obviously important that we need to minimize operations and maintenance (O&M) costs." *(Maintenance and Reliability Best Practices*, p. 177)

And further "...the major portion of O&M cost becomes fixed during early design and development phase of the asset. There are ample opportunities to reduce the Life Cycle Costs (LCC) during the design, building, and installation of the asset." (p. 177)

Gulati's discussion on this subject ended with this note, "Assets should be designed so that they can be operated and maintained easily with minimum operations and maintenance needs." (p. 178)

For our exercise let's assume that our equipment is complete with the components we have listed, and we want to improve the asset's reliability. It's been stated here in this section that you can't maintain more reliability into a machine than it has inherently. The only course of action to increase the reliability of a machine is to take measures to do just that—increase the reliability.

This may be the easiest question you'll be asked. If we want to increase the reliability of an asset, do we modify it with components of higher or lesser reliability? Higher, right? For our discussion, we might consider some of those components that are listed to the far right in Table 4-1.

We do need to exercise some caution when modifying our equipment. The boss will ask us to make the machine run faster by putting a smaller sprocket and a bigger chain on it. However, we need to consider the additional stress that we are putting on the asset. I often ask people to consider that a more powerful cylinder will lift a heavier load, but it also applies force in the other direction—against the machine frame, or the concrete mounts. We may also aggravate a vibration balance, or add additional heat. Our PM strategy will definitely need to be modified. We are walking a very thin line on reliability with *blind* modifications for more speed.

Our strategy should be to build the machine with the highest reliability and then maintain that!

OEE

Overall Equipment Effectiveness (OEE) is easily the most abused, confused, and lied about metric in the *entire world*. I added that last bit for dramatic effect. But if we're being honest, isn't there a lot about this metric that we don't know, or trust?

I came across a maintenance manager who bragged to me that his company's OEE was 137%. I was not a math major in college but I did take a lot of math. If we are calculating this measure correctly, we cannot exceed 100%. It is mathematically impossible. Getting to 137% would require a lot of rounding up.

What is OEE and why do we care?

Does Our Boss Even Know What OEE Is?

There are many wonderful texts and other media explanations as to what Overall Equipment Effectiveness is. What never seems to be mentioned, however, is that OEE can be whatever we say it is. It is true that most consider OEE as a measurement of the six big losses in the three major categories. Table 4-2 provides a fresh reminder of these three categories and subsequent losses.

Table 4-2: Overall Equipment Effectiveness

	Availability
Breakdowns	The equipment fails to perform one or more of its functions
Changeovers	The equipment is taken off line for the purpose of changing product
	Performance
Jams	The equipment is blocked by raw material or machine movement
Minor Stoppages	The equipment either stops momentarily on its own, or is stopped on purpose for any minor issue
	Quality
Yield Loss	Machine start up loss before good product begins to flow
Scrap	Off specification product

Table 4-2 is what many consider to be the classic OEE definition, but it can be so much more than that. Regardless of our definition we have to determine what value there is to calculating OEE and how we can improve on that value.

OEE is a great tool to use when appropriate. It is not a very good tool to use all the time. This metric is a great resource that can be used as a type of metric matrix, combining the required deliverables of maintenance and operations. It should come as no surprise that until maintenance owns part of the production deliverables, they (maintenance) will never care about those measures as much as production. Isn't the inverse also true? Production has to own part of the maintenance requirements for there to be interest on their (production's) side.

OEE provides this *matrix* of *metrics*.

We Most Likely Don't Have a Real *Definition for OEE*

Unfortunately, there is likely to be little in the way of explanation as to what each category and sub-element actually means.

For example, Availability is parsed into changeovers and breakdowns. What exactly is a changeover? Is it a complete product change, or a slight modification? What is a breakdown compared to the Performance sub-element of minor stoppage?

When we get lost in the minutia we tend to fade the lines of exactly where the problem lies. OEE was meant to tell us in what category the problems lie, and in which sub-element corrections need to be made.

Availability is a unique category for the primary reason that if we suffer one of the issues in this area there is no product being made. In the Performance and Quality categories, when issues arise product is still being made. Availability is quite literally a real showstopper.

Interestingly, most maintenance and reliability individuals can recount at least one instance in their careers where their facility suffered a major failure, a breakdown. Of course there was the usual frustration, finger-pointing, and crisis management, but eventually the machine was brought back on line, and production continued. The line was down, production lost, and there were many bruised feelings and egos.

Juxtapose that to a changeover, conducted by operators, that takes twice as long as it is scheduled to take What rending of clothing and gnashing of teeth results from this clear violation on the part of production? Probably very little.

But isn't the result of both scenarios the same? No product?

It might be that we are weaponizing OEE and not really using it for the good in which it was intended. Our error might be founded in our blind belief that our equipment is capable of running and performing whenever we desire.

Let's look again at the categories that make up the classic OEE. There is absolutely nothing wrong with this approach. In fact, I've used it many times myself. What does beg to be answered is the question "Does anything else factor into the effectiveness of equipment?"

OEE In Contrast to Other Measures and Means

Perhaps a short side-by-side look will help. Consider Table 4-3, which lists the more common aspects of some of the concepts we've been discussing.

Table 4-3: Comparison of common concepts

Inherent Reliability	Eqi pment Reliability	OEE
How it's designed	Does what it's supposed to do	Quality
How it's installed	When it's supposed to do it	Availability
How it's built	For as long as it's supposed to do it	Performance

Looking at Table 4-3 again, ask yourself a very simple question: Does the Availability measure (breakdowns and changeovers) have any root in either inherent reliability or equipment reliability, as defined by our loose street definition? Isn't it likely that we are measuring the OEE on an asset whose reliability has been compromised by endless modifications, as well as operator and maintenance abuse?

Couldn't it also be true that the equipment was originally designed to do what it's supposed to do, when it's supposed to do it, for as long as it's supposed to do it, but we blew our chances in the design and build stages of the asset?

How would those factors be recorded in an OEE formula? The problem is, they are not recorded or even considered when we establish an arbitrary OEE goal.

There is a popular belief that the world-class value of OEE begins at 85%. The most popular example of this would be an asset whose Availability is 90%, Performance is 95%, and Quality is 99%. It would be very difficult to hit these numbers, on average, for an entire year.

$$.90 \times .95 \times .99 = 85\%$$

This might be an example formula that you have seen before.

If the world-class OEE number is not 85%, as most have come to believe it is, then what is it? Is it even a metric that has a world-class equivalent? Given the cacophony of industries, assets, components, operators, engineers, designs, etc., wouldn't establishing one world-class OEE number be an apples-to-oranges comparison?

The argument that we are making is that the ability to consistently hit an Availability of 90% might have been determined for us back in the design, build and install phases. Our original design was for an asset that could do what we wanted it to do, for as long as we needed to do it. Compromises on the project budget, timelines, and component selection sealed our fate. Did we ever change our anticipated result? The original equipment design looked good on paper.

Companies have spent time and money to figure out how to collect and report OEE on a continuous and running basis. They spend more time and money chasing a value (say, 85%) that might not be feasible based on the design, build, and install phases of Equipment Design Excellence. The result calls to mind the dog chasing its tail image, and continues bruising feelings and egos.

Instead, why don't we consider another, more focused use for OEE?

Consider an approach that uses Overall Equipment Effectiveness as a measured method to pinpoint an asset's OEE value, at a specific moment in time. Also, subsequent application of the metric is used to capture a *delta*, a change, in the asset's overall contribution to our production or service efforts.

Here is an example. One month prior to a modification or improvement effort on a hydraulic press, continually measure the classic OEE values as we've come to understand them: Availability, Performance, and Quality.

Perform the modification or improvement when scheduled. Next, measure OEE for another two months, as before. Make a comparison of the values, and record the change in each category and sub-element. Lastly, record and report on the delta.

This is not only a unique and well-suited use for this valuable tool, but also helps to formulate our compelling case for more improvement projects in the future. The calculation and explanation of a positive change in availability, performance, or quality can go a long way in quantifying the return on investment.

Figure 4-2 on the following page, shows the OEE calculation for the hydraulic press we discussed earlier. These are the pre-modification numbers collected over one month prior to the upgrade work. This is a one shift operation, Monday through Friday, for a total of twenty operating days for the month of January. Calculate the Availability, Performance and Quality values, and the OEE.

Reminder—enter your responses in the greyed cells.

After the modification, the press OEE was calculated for the two months immediately following the improvement. Again, this is a one shift operation, Monday through Friday for 40 operating days. Figure 4-3 is a record of those new values.

Calculate the Availability, Performance, and Quality values and the OEE.

This press is used to press the initial form of a stainless steel sauce pan. Each additional percentage of availability equates to an additional $140,000 to the company's profit. Each additional percentage of performance equates to an additional $50,000 to the company's profit. Each additional percentage of quality equates to $40,000 to the company's profit. What is the delta in the OEE

Machine:	HP Press	Date:		OEE %
Product:	Sauce Pan	Shift:	Jan 1-Jan 31 1st	
AVAILABILITY				
1. Gross Available Time:	If 8-hour shift=480 min., if 12-hour shift=720 min			9,600 Minutes
2. Planned Downtime:	Time for meetings, cleaning, breaks, PM, and other scheduled events			1,100 Minutes
3. Run Time:	Gross Available Time – Planned Downtime	#1 - #2 =		8,500 Minutes
4. Downtime Losses:	a. Breakdown minutes <u>600</u> b. Changeover minutes <u>600</u> c. Minor Stoppages <u>200</u>	a + b + c =		1,400 Minutes
5. Operating Time:	Run Time – Downtime Losses	#3 - #4 =		7,100 Minutes
6. AVAILABILITY:	Operating Time÷ Run Time x 100	#5 ÷ #3 x 100 =		%
PERFORMANCE				
7. Total output during operating time:	Parts, yards, units run per shift: good & bad			Total Units 248,500
8. Theoretical Cycle Time (ideal):	Minutes per unit			Min./ Units 0.022
9. PERFORMANCE:	[(Theoretical Cycle Time x Total output) ? Operation Time]	[(#8 x #7) ÷ #5] x 100 =		%
QUALITY				
10. Rejects during operating time:	(machine related defects)			24,850
11. QUALITY	(Total output - Rejects /Total Output. X100)	[(#7- #10)÷ #7] x 100 =		%
OEE				
12. OVERALL EQUIPMENT EFFECTIVENESS	(Availability x Performance x Quality x100)			%

Figure 4-2:
Pre-Modification
OEE

Machine:	HP Press	Date:	1 Feb through 31 Mar	OEE %	
Product:	Sauce Pan	Shift:	1st		
AVAILABILITY					
1. Gross Available Time:	If 8-hour shift=480 min., if 12-hour shift=720 min				19,200 Minutes
2. Planned Downtime:	Time for meetings, cleaning, breaks, PM, and other scheduled events				2,200 Minutes
3. Run Time:	Gross Available Time – Planned Downtime	#1 - #2 =			17,000 Minutes
4. Downtime Losses:	a. Breakdown minutes 250 b. Changeover minutes 400 c. Minor Stoppages 50	a + b + c =			700 Minutes
5. Operating Time:	Run Time – Downtime Losses	#3 - #4 =			16,300 Minutes
6. AVAILABILITY:	Operating Time÷ Run Time x 100	#5 ÷ #3 x 100 =			%
PERFORMANCE					
7. Total output during operating time:	Parts, yards, units run per shift: good & bad				Total Units 665,000
8. Theoretical Cycle Time (ideal):	Minutes per unit				Min./ Units 0.022
9. PERFORMANCE:	[(Theoretical Cycle Time x Total output) ÷ Operation Time]	[(#8 x #7) ÷ #5] x 100 =			%
QUALITY					
10. Rejects during operating time:	(machine related defects)				56,500
11. QUALITY	(Total output - Rejects /Total Output. X100)	[(#7- #10)÷ #7] x 100 =			%
OEE					
12. OVERALL EQUIPMENT EFFECTIVENESS	(Availability x Performance x Quality x100)				%

Figure 4-3:
Post-Modification
OEE

Machine:		Date:		OEE %	
Product:		**Shift:**			
AVAILABILITY					
1. Gross Available Time:	If 8-hour shift=480 min., if 12-hour shift=720 min				Minutes
2. Planned Downtime:	Time for meetings, cleaning, breaks, PM, and other scheduled events				Minutes
3. Run Time:	Gross Available Time – Planned Downtime	#1 - #2 =			Minutes
4. Downtime Losses:	a. Breakdown minutes _____ b. Changeover minutes _____ c. Minor Stoppages	a + b + c =			Minutes
5. Operating Time:	Run Time – Downtime Losses	#3 - #4 =			Minutes
6. **AVAILABILITY:**	Operating Time÷ Run Time x 100	#5 ÷ #3 x 100 =			%
PERFORMANCE					
7. Total output during operating time:	Parts, yards, units run per shift: good & bad				Total Units
8. Theoretical Cycle Time (ideal):	Minutes per unit				Min./ Units
9. **PERFORMANCE:**	[(Theoretical Cycle Time x Total output) ÷ Operation Time]	[(#8 x #7) ÷ #5] x 100 =			%
QUALITY					
10. **Rejects** during operating time:	(machine related defects)				
11. **QUALITY**	(Total output - Rejects /Total Output. X100)	[(#7- #10)÷ #7] x 100 =			%
OEE					
12. **OVERALL EQUIPMENT EFFECTIVENESS**	(Availability x Performance x Quality x100)				%

Figure 4-4:
Blank OEE form

before and after the modification; what value was delivered to the company?

Pre-Modification OEE: _____
Post-Modification OEE: _____
Delta in OEE: _____
Value (financial gain): _____

If you feel brave enough, Figure 4-4 is a blank form. Take this form out to one asset and conduct a one week OEE measurement.

Maintenance and Reliability Strategy

You may recall that we developed a draft strategy earlier in this workbook. In Chapter 2 we laid out the basic strategy for the Normandy invasion in World War II and then we superimposed a maintenance strategy over that. In this section we are going to delve further into the ideas of maintenance and reliability and work together to draft a solid approach to improving both.

We start by restating your definitions of reliability and maintainability. It is all right if your understanding, and therefore your definitions have changed over the pages of this book. Record your original, or perhaps more modified, definitions for the terms that follow:

Maintainability:

_____

_____

_____

Reliability:

_____

_____

_____

Again, it is perfectly fine if your beliefs have changed over the course of our discussions and you have a new definition. We will go forward with what you have recorded here.

Synonymous

Maintainability has come to be synonymous with *Mean Time To Repair* (MTTR). Reliability has become synonymous with *Mean Time Between Failure* (MTBF). These two connections have given us an ability to really develop a further strategy toward both maintenance and reliability.

Breaking Down The Breakdowns

If maintainability is equivalent to MTTR, and we understand MTTR to mean the time between a work stoppage and the asset being brought back on line, we can sketch that process to look something like Figure 4-5.

MTTR

| Response time | Troubleshooting time | Repair time | Startup Time |

Figure 4-5: MTTR

A real study of Figure 4-5 leads us to believe that we can improve maintainability by shortening MTTR. For this metric, the smaller the number, the better for us.

Give Figure 4-5 a hard look. Response, troubleshoot, repair and aid in startup. Doesn't that sound a lot like what maintenance does?

Our *strategy* for maintenance, only as it regards MTTR, now becomes the methods we can apply to decrease the time it takes to perform each of these individual functions.

What Do You Think?

In the space provided, record your thoughts on how each of these areas can be reduced as far as the time necessary to complete. I've provided some examples to get you started:

Response time:
Establish a Do It Now (DIN) crew that responds directly to trouble calls:_____

Troubleshooting time:
Enhance efforts toward building objective-based and skill-based training, concentrating on troubleshooting:

_____

_____

Repair time:
Establish a parts standardization program:

_____

_____

_____

Startup time:
Company policy that all operators stay with their equipment during breakdowns (unless the down time will be > 1 hour):

_____

_____

_____

It is very important to note that we've only described examples of how we might shorten MTTR. Maintainability and the maintenance of our equipment go way beyond this simple approach.

It Starts With a Strategy

Our maintenance strategy has to include improvements to our equipment before it is even built. The best way to accomplish this task is to include maintenance technicians and operators in the design phase of an asset or the modification of an asset. The equipment has to be designed and built to be maintained.

I have seen weekly PM tasks written for components that are physically off the ground and located behind the machine. Sometimes these machines and their parts are shoved to the back of a facility in a corner. I'm sure that when the engineer attended the run-off meeting, all the components were in plain view and quite accessible.

Design In the Maintainability

List asset design characteristics that would aid in the maintenance (think *access*) of an asset. I've listed a few to get your creative juices flowing.

- Access doors
  _____

- Remote test ports
  _____

- Quick disconnects
  _____

It's not only the design, build, and install phases in equipment design excellence that need to be addressed to achieve a high likelihood of maintenance excellence, it is also the manner in which we choose to organize our maintenance resources. Ideas around organizational structure will be addressed in a later section in this chapter. There is one organizational element that dovetails nicely into this immediate topic—DIN crews.

A Hit Squad

In the example just provided, I suggested establishing a DIN, or *Do It Now* crew. This particular group of highly skilled and capable technicians responds to all breakdown calls. When there are no breakdowns to react to, the DIN crew members busy themselves with completing low priority planned and scheduled work.

Some maintenance organizations aren't resourced to a level that will allow them to have a separate DIN crew. That is unfortunate, as this crew also serves to run interference for the preventive maintenance effort. You might agree that it is the breakdown, or trouble calls, that impede our ability to get PMs or other scheduled work accomplished.

If you could form a Do It Now crew, what skills would be represented within this group? List those talents here:

_____

_____

_____

How would you suggest they be equipped?

Our maintenance strategy isn't complete unless we also address access to spare parts. Spare parts begin and end with an asset's *Bill of Materials* (BOMs). Until we have total commitment to demanding, obtaining, and maintaining the integrity of the BOMs, we will never (ever) nail this element of maintenance and maintainability down. There will be further discussion on BOMs in Chapter 6.

Putting It Together

Our maintenance strategy now consists of the following high points:

- Design, build, and install an asset with the maintainability of that asset in mind

- Demand, obtain, and maintain the BOMs

- Work to reduce the MTTR

This is a very short list that begins the discussion of a maintenance strategy. Ultimately, the organization needs to agree in principle as to what level of care it is going to allow. We can maintain a machine in showroom condition, or run it to destruction. I suggest that one of those levels may create a lot of unnecessary work.

Based on our very young discussion on maintenance strategy, what additional elements would your strategy include? List those here:

_____

_____

_____

If we are going to conceive of a strategy to maintain an asset, we must also ensure that we have a strategy to have an asset that is reliable to start with.

It should not escape the attention of anyone who has made it this far into this workbook that we have been developing a maintenance and reliability strategy this whole time. In regard to a pure reliability strategy, there are many elements that must become second nature.

The Space Between

First, review the model created for the discussion of maintainability in Figure 4-5. Figure 4-6 is a similar representation but for Mean
Time Between Failure.

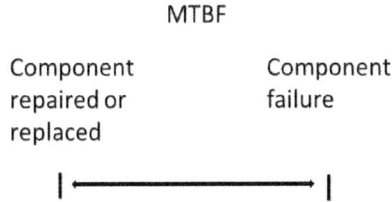

MTBF

Component	Component
repaired or	failure
replaced	

Figure 4-6: MTBF

It is a sure bet that you've never seen MTBF graphically
depicted as in Figure 4-6. Most diagrams include MTTR and other
elements within the MTBF graph. I chose to cut to the chase here
and talk about what we are really interested in, which is how long
an asset or component remains viable.

Reliability is synonymous with MTBF. As such, our task is to
increase this time span. We will explore the means by which this
is made possible, but we shouldn't abandon the Mean Time To
Repair completely. Keep in mind that we are ultimately concerned
with how much time the component is in service.

In the Beginning

We are going to start back in the design, build, and install phase of
our asset. Actually, look back at Figure 4-6 and substitute *Asset* for
Component. The same desires hold true. Regarding the infantile
stage of our asset, and thus our component selection in the beginning, we can make some tremendous inroads toward a reliability
strategy by requiring *Original Equipment Manufacturers* (OEMs)
to use robust and proven components in their design. We have
previously discussed the idea of demanding some standardization
in our parts selection for new construction and equipment modification. Reliability really does start here.

Round Peg, Square Hole

Any item chosen apart from that which we know to be the best part for the job could be a slight against our efforts to build a reliable asset. Beyond the design and build segments is the installation portion of a project. Consider the dangers to reliability that might hover around this phase.

I'm going to start a list of installation hazards to avoid. See if you can add to it:

Not isolating the asset or component from vibration

Environmental agents: heat, cold, moisture, dust

_____

_____

_____

_____

Extending the time intervals in the reliability equation goes beyond what we can accomplish in the beginning phases of an asset's development. The way we care and maintain the equipment also affects the reliability of the asset and the components on the machine.

The Value of Training Gets Lost in the Fog

As stated earlier, we can't make an asset more reliable unless we modify it with that end in mind. Said another way, we can't maintain more reliability in a machine than it has inherently. If that is the case, and it is, we need to ensure that everyone takes a level of responsibility in the care and upkeep of the capital equipment.

A major element of our reliability strategy is to make sure we have equipped our workforce to be able to guard the reliability our equipment arrives with. This is accomplished with a training program that delivers: theory of operation, maintenance, troubleshooting, lubrication actions, and cleaning. This training must be objective-based. Objective-based training requires learners to demonstrate their knowledge of the practice and their ability to perform that task.

Right Part, Right Time

Our reliability strategy continues with assurances that we replace components with like-for-like components. It should also assure that we certify any repairs or rebuilds made to the components, either in-house or at an off-site repair shop.

TLC—An Example

We just touched lightly on the understanding of cleaning and lubrication. I would expect little debate that keeping equipment clean, lubricated, and all the fasteners tightened will lead to an asset that is more reliable. At a minimum, this type of shop floor care will return an asset that is more pleasant to operate and work on.

During my time in the Air Force, I made a previously unknown connection to the cleanliness of an aircraft and its performance.

I was stationed at a B-52G bomber and KC-135A tanker base in Michigan in the late 1980's. During my time there, I made close friends with a fellow 2nd Lieutenant maintenance officer who was in charge of the back shop branch that owned the aircraft wash hangar.

The aircraft wash hangar was the largest hangar I had ever seen up to that time. It could hold an entire B-52G, including the tail section. I realize this does not provide any point of reference to those who do not know what a B-52 is, but believe me, it is a big plane!

I was coordinating aircraft wash schedules with my friend and we were discussing aircraft tail numbers in particular. It was during this discussion that I started a very innocent record of aircraft (bombers and tankers) and their time in the wash rack and their next scheduled sortie (flight).

After a short time, there formed a pattern that basically showed a connection between the cleanliness of our aircraft and its flight performance on its very next flight. The evidence showed, and it could be proven, that a clean plane was perceived to 'fly better' by the aircrew. In the case of my informal study, a plane was likely to receive fewer inflight discrepancies on the very first flight right after it was washed.

I know this isn't science, but it seems to be true. Doesn't your car have a different *feel* after it has been washed and waxed? Doesn't it also seem to run and perform better?

A reliability strategy requires that we do the right things right.

Putting it Together

Our reliability strategy to this point has these key elements:

- Design, build, and install an asset with the reliability of that asset in mind

- Replace components with like components

- Modifications should increase reliability

- Keep the asset clean, lubricated, and all fasteners properly tightened

- Work to increase the MTBF

The engineering department can set the organization up for success by demanding components on new or modified equipment to be like the components that seem to already be working in our crazy environment. This must be insisted upon.

What additional elements would your reliability strategy include? List those here:

_____

_____

_____

Cost Benefit Analysis

It's certainly one thing to have a maintenance and reliability strategy and quite another thing to pay for it. Compelling others to see the value in what you have assembled as your vision and mission may be wrought with objections and outright criticism. My age-old nemesis in manufacturing was always the plant controller, or the occasional purchasing clerk who would blankly ask me "Why do you need to buy that?"

I will admit that I wasn't always prepared for mental combat with these agents of finance. In fact I often rode into a battle of wits against the front office staff completely unarmed. I know better now, and so will you. I knew what I needed, but I just wasn't speaking their language.

Putting This in Terms We Can All Understand

Building a cost benefit analysis, at least informally, is almost an automatic reflex for most adults. When we were children, we had no sense of the long term effects of our decisions, nor did we really care. In fact, most people can remember standing in front of a candy aisle with one dollar held tightly in their hand, faced with an almost impossible decision—what to choose? In our little hearts we just knew that we'd never have another dollar to spend on candy. This was our one shot.

Fast forward thirty-five years and that same kid is looking over a car that his own child is considering buying. Would it be fair to say, in a very loosely defined way, that in both cases that original kid is looking for value for the money in the same way?

Consider the list of attributes of a car that you want to buy as shown in Table 4-4. You have the option to buy a used car that is in good condition, or the very same car, just a newer model, that is brand new. Starting from the number one, in both categories, number the attributes in order of importance for you, with the number one being the most important.

Table 4-4: Car comparison

Attributes	Used	New
Low monthly payments		
Total costs (tag, title, tax)		
Mileage (odometer-reading)		
Gas mileage		
Condition		
Overall cost of ownership		
Safety rating		
Age		

Almost no one chooses overall cost of ownership as one of the top five most important attributes. Ask a co-worker to complete this same list and see what views you may have in common.

The Meaning of 'and'

There are two different ideological and different meaning elements to a cost-benefit analysis: cost and benefit. One error we often commit in maintenance and reliability is to assume that everyone looks at the cost and the benefit in the same way.

If you recall our discussion in Chapter 1 regarding critical thinking, we can be certain that personal history, education and job requirements are going to tilt the prism through which we look at a particular situation. A finance or purchasing associate is going to look at cost in a way that is different than a maintenance technician or a front line supervisor.

We should celebrate and take great comfort in knowing that there is an organization within the company that looks out for the money. Someone has to be fiscally sober while the rest of us are metaphorically on spring break with dad's credit card.

I think there is one simple comparison that brings the complexity of analyzing the payoff or payback of cost vs. benefit into sharper focus, and that is cheap vs. inexpensive.

Would You Prefer Something That Was Cheap or Something That Was Inexpensive?

Cheap is a term that should conjure up an image of poor quality. Something that is cheap is something that is poorly designed, a poor fit for the purpose, made of poor quality material, and is generally garbage, for lack of a more poetic analogy. Some make the mistake of believing that getting an item cheaply is the same as paying a cheap price for a quality item. It is not.

An inexpensive item seems to reflect an item of quality that can be purchased or obtained for a bargain. You are getting good quality at a good price.

Have you ever said in reference to the price for something, "That looks too good to be true?" Or, "What's the catch?"

The exercise now becomes to convince others to see that the request being presented is one that is inexpensive (good value, low price) when they may be hard-wired to angle for the option that is cheap.

Savings Despite the Cost—A Real World Example

Years ago, I was working with a client to resolve an issue using some *Root Cause Analysis* techniques. This was an oil field operation (similar to the one described in the section on RCA). The problem was that the stuffing boxes on the oil wells were leaking.

The stuffing box is a device that is 'stuffed' with a greased wool, cotton, or other fabric-like material. This material seals around the shaft or pipe and prevents leaks as the shaft moves. There may also be other primary and secondary seals in the stuffing box. A good example is to think about a boat's propeller. The propeller is in the water and the drive shaft passes through a stuffing box to connect to the drive side which is inside the boat. No water leaks from the lake side to the boat side because of the stuffing box.

The situation that the client was facing was that all their wells were beginning to leak around the stuffing boxes as the polished rods moved through them. This is wholly undesirable, as it can create an environmental hazard and it also leads to a loss of product and production.

In our Root Cause Analysis we discovered that the purchasing department had changed the type of seals that were bought for the rebuild of the stuffing boxes. The new seals were not effective against the sandy oil that was being extracted. In other words, they wore out quickly.

When interviewed as to why they (purchasing) would make a unilateral decision to change these seals, their response was classic, "They were cheaper." Yes, they were.

Cheap is not better, but then again, neither is expensive.

That purchasing agent had saved the company hundreds of dollars by researching and getting a better deal on a slightly lower quality seal. In doing so, that agent cost the company tens of thousands of dollars. That would be an example of a negative cost-benefit analysis.

There are no formulas in this section on cost-benefit analysis, just plain speak. We know from earlier discussions that in order to have an asset, that from the start is maintainable and reliable, it must be designed, built, and installed for those functionalities. From there the job involves maintaining the asset properly and performing the actions to guard the inherent reliability of the equipment.

As such, we need to focus not on actions that are more expensive, or parts and services that cost more, but rather the types of activities that make our equipment maintainable (lower the MTTR) and shore up the reliability, if not improving the reliability through modifications (MTBF).

Read the following scenario and provide your suggested direction where indicated.

Exercise—Your Turn

A company located in the Midwest operates three milling machines. Each machine is essentially the same; just the serial numbers are different. One of the more critical parts on the milling machine is the main drive gearbox. The gearbox on each machine is maintained by a series of inspections and service actions. The operators keep them clean, maintenance keeps them running true and secured tightly, and the lubrication tech provides periodic oil sampling, flush, and service when needed. This is a robust device, and has served well on all three mills.

The main drive gearboxes, one per mill, have performed so well, and have lasted so long, that there has never been a need to replace them. Each has been in service for ten years.

During a routine pre-shift walk around, the operator on mill #2 discovers an oil leak from the gearbox. Subsequent inspection by maintenance indicates a crack in the cast housing. There is a temporary fix, a patch, but it is absolutely necessary to replace the gearbox. Once replaced, the old gearbox cannot be permanently repaired. It will have to be discarded. Inspections on the other two gearboxes come back negative; there is no damage.

While researching where to purchase a new gearbox, just like the one that is on the mill, it is discovered that this item is no longer available. No OEM supports this device. There are a few to be purchased from surplus dealers, but this item is obsolete. The cost from a surplus dealer is $15,000.

Procurement has, however, found a source for another gearbox, a different make and model, but the functions are exactly the same. This new gearbox is 'drop in ready,' with the same gearing, same bolt on pattern, and the same shaft sizes. The only troubling aspect is that it looks cheap.

Engineering has found yet another gearbox that will work, and it is also 'drop in ready,' and is widely available and looks stouter than the original. The trouble with this choice, quite frankly, is that it is $20,000 vs. the one sourced by procurement for $10,000.

Engineering has also suggested that since the old gearboxes are obsolete, perhaps an addendum needs to be a created on the ten-year plan to retrofit the other two mills with this new gearbox.

To recap our choices, this company can either:

- Get the same gearbox, used, from a surplus dealer for $15,000

- Buy a visually cheaper looking gearbox for $10,000

- Buy a visually stouter gearbox for $20,000 and plan to retrofit the other two mills

Time is not on their side.
What would you advise, and why:

_____

_____

_____

_____

Maintainability and reliability come at a cost. It was mentioned earlier that it is a good thing that someone in the organization is looking out for the money, because that allows the company *to do business*. Thank goodness, also, that someone is looking out for the maintainability and reliability of the equipment, because that keeps the company *in business*.

Our cost-benefit analysis becomes very simple if we agree on some of the founding principles of basic reliability.

- We should always demand standardized parts on new construction or modifications—those parts that have been proven robust enough to work in our environment

- We should always insist on like-for-like replacement at all times when practical to do so

- When modifying equipment, we should endeavor to improve reliability with more robust components, not just components that make the machine run faster

- We should change our spare parts and PM strategies as the machines are changed and get older

All of this costs money. We do not benefit from doing it on the cheap, but we don't have to blow out the checkbook either.

Stated a few pages earlier, there are two different ideological and different meaning elements to a cost-benefit analysis: cost and benefit. We've discussed cost to some extent and we've touched on benefit as well.

Recall your suggested advice to the company and its mill gearbox predicament. It is hard to debate cost without also having to explain the benefit that can be derived from the decision.

Look at these choices one more time:

- Get the same gearbox from a surplus dealer for $15,000

- Buy a visually cheaper looking gearbox for $10,000

- Buy a visually stouter gearbox for $20,000 and plan to retrofit the other two mills

The actual costs are listed in the options themselves. What are the benefits? Note: You most likely listed the benefit to your choice with the selection you made earlier.

Table 4-5 lists some benefits for each choice.

Table 4-5: Cost vs. Benefit

Option	Cost	Benefit
Surplus dealer	$15,000	Like-for-Like, delivery time, availability
Procurement	$10,000	Less expensive, delivery time, availability
Engineering	$20,000	Delivery time, availability, long-term solution, improves reliability

If we cancel out the benefits that are common among all three, we are left with Table 4-6.

Table 4-6: Narrowing down the choice

Option	Cost	Benefit
Surplus dealer	$15,000	Like-for-Like,
Procurement	$10,000	Less expensive
Engineering	$20,000	Long-term solution, improves reliability

Which one of these choices delivers a pathway to sustained reliability and maintainability?

Sometimes in our attempt to compel other forces, we forget that our role is to ensure capacity and equipment availability. I'm not negating the importance of being fiscally responsible, but what other party in plant or corporate leadership is arguing for reliability and maintainability of the assets? We can't all make the argument for cheaper when what we need is less expensive.

Organizational Structure

It would seem only fitting that at some point we discuss the actual execution of maintenance, and that is where this section fits nicely into the buildup. At this point, we have done our work to ensure a high level of reliability. We have considered the equipment design, build, and installation, within budget and with maintainability and reliability in mind. The equipment is ready to work. Are we?

My Sad Story

I was exactly three months out of the military, in my new civilian job as a plant engineer and maintenance manager in a small

Illinois town manufacturing facility. I had a twenty-nine inch waist and hair I actually had to comb. I had been a senior captain, aircraft maintenance officer, experienced crash investigator, and distinguished graduate of every single Air Force course I ever attended, including the Maintenance Officers Course. I knew maintenance inside and out.

I arrived at a plant that had no:

- Preventive Maintenance
- Work order system
- Predictive Maintenance
- Computers
- Planning or scheduling
- Training program
- Storeroom (not even a closet)

I was going to teach these forty-nine technicians how to run maintenance and how to be organized and professional about it.

I realized quickly that I really didn't know how to run a maintenance organization. I had never actually run one. My Senior *Non-Commissioned Officers* (NCOs) did all that for me. I also didn't know how to start a maintenance organization, as the Air Force already had one when I joined. They just told me not to screw theirs up. That left me in a very bad predicament. I may have oversold my resume a little; certainly my qualifications.

I received a flyer in the mail one day. We didn't have computers in maintenance as I mentioned earlier, so we were still getting those laminated line-cards that salespeople mail out. I got one and gave it a serious study. It listed, almost as I detailed out in the previous bulleted listing, what I was experiencing. The flyer went on to mention that the publishers of this particularly timely document could be my salvation, if I'd only call 1-919... And I did.

The owner of that company came out to visit me a few weeks later and assessed that I was as near 100% reactive as he had ever seen. We spent a little time strategizing and compelling my boss to let go of a few dollars to hire a consultant to come down from Chicago, IL to help me put together a maintenance program and maintenance department.

That began my relationship with Marshall Institute.

We Got To Work

The consultant they sent down from Chicago was a fireplug of a man named Jack. Jack was as thick and no-nonsense as a person could be. He had no neck, I swear. He wore a permanent scowl, a bushy mustache and an ill-fitting suit coat. Jack looked like he meant business.

I only wish that Jack would have shown up on day one carrying a violin case. Back then, when you told someone in Illinois that a person was coming down from Chicago to take care of business, it meant something.

Jack was very much a believer that if you want something different, you have to do something different. Years before Roger Lee connected the idea of a proper organizational structure, Jack was moving me in that direction.

"A proper team structure is required to maximize your efforts when implementing a cultural change to any organization. You are just feeding your maintenance insanity when you leave the same structure in place and the same people in the same roles forever." (R. Lee, *The Maintenance Insanity Cure: Practical Solutions to Improve Maintenance Work*, p. 134)

Jack and I worked together for months and put together a dynamic maintenance organization that moved us from reactive to 98% consistent uptime in eighteen months. Here's what Jack taught me.

Teaching a New Dog Old Tricks

Maintenance has to perform three functions at the same time, all the time:

1. Preventive maintenance

2. Corrective maintenance

3. Emergency maintenance

Each of these has to be performed without interference from the other.

Have you ever had to pull a technician off of a preventive maintenance task to handle a breakdown? Have you ever had to pull a craftsman from a planned and scheduled job to address an emergency?

We'll learn in Chapter 6 why preventive maintenance needs to be protected. As far as corrective maintenance goes, we've made a contract with production. Our contract with production is the maintenance schedule. We need to be on the job until we are done. We don't want to break a contract.

Jack said we were going to bring in some craftsmen and develop a maintenance philosophy, vision, and mission. After that, we were going to design a maintenance organization that could respond to all three activities without interruption.

I had forty-nine maintenance technicians, mechanics and electricians. Of those, fifteen were tool and die makers, so they were carved out and worked strictly in the tool and die shop.

Let's take a quick break from my story to document your resources. In Table 4-7, record your current maintenance staffing. List the skilled trade categories along the top, and identify your different shifts along the left column.

Table 4-7: Maintenance staffing

Trades _ _						
Shift:						
Shift:						
Shift:						
Shift:						

I had thirty-four electricians and mechanics to organize so that we could perform all three types of maintenance: preventive, corrective, emergency.

Jack instructed us to first create a Preventive Maintenance team. After developing the roles and responsibilities for a PM team member, we recruited from within our ranks and then formed, chartered and trained the folks to be a day-shift only, PM crew.

Since we had no preventive maintenance at all to begin with, this team developed every single PM task and worked to improve and refine those tasks until the program was humming along nicely. The PM crew also included the oiler who we already had on staff. The PM crew had a lead man who ran the show.

This took care of the preventive maintenance that had to be performed.

For emergency maintenance, Jack led us through the development of three, three-person hit crews. I've come to understand that they are called Do It Now squads today, but back then we called them hit crews.

There was one hit crew per shift whose role was to respond to breakdowns. Each crew had a lead man, and the lead man would determine who would stay and repair the equipment based on what the failure was.

When the hit crew was not addressing breakdowns or other emergency issues, they were concentrating on low priority work. Low priority work are those approved work orders that seem to float to the bottom of the work order system and never seem to get completed.

The remaining mechanics and electricians were assigned to various production lines as line mechanics or line electricians. They moved their tool boxes out to their assigned lines and set up functioning work spaces out in production. Their job was to work on corrective maintenance assignments and to keep the lines running.

The Structural Evolution (You Know I Wanted to Write 'Revolution')

Our day shift was the second shift. Our initial organizational structure was something like Figure 4-7 on the following page.

I was fortunate that my boss was very supportive, and due to the initial success of our organizational structure, I was able to reach out and create a storeroom from scratch and hire a maintenance manager, reliability engineer, and planner.

One success we had that I was unprepared for was that we stopped having emergency calls on day shift (our second shift). If the operators were having a particular issue with a machine, they simply called the line technician over and the problem was corrected quickly. This isn't necessarily ideal today, but when we

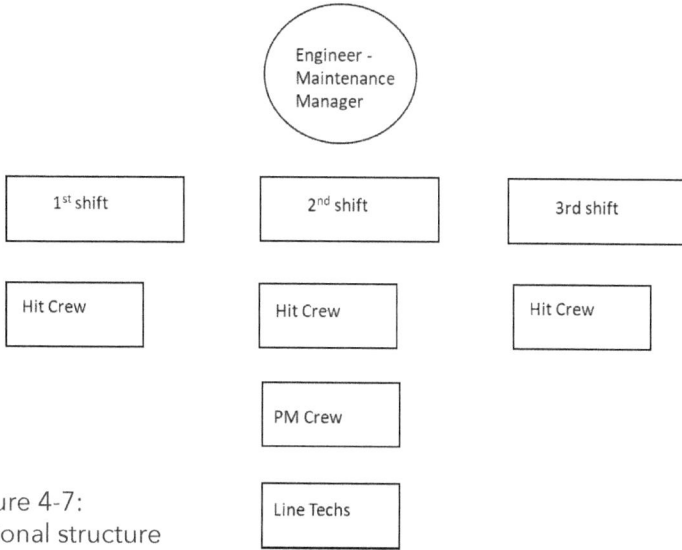

Figure 4-7:
Organizational structure

were just trying to bring some structure to maintenance, it was very helpful.

Our final organization was similar to that reflected in Figure 4-8.

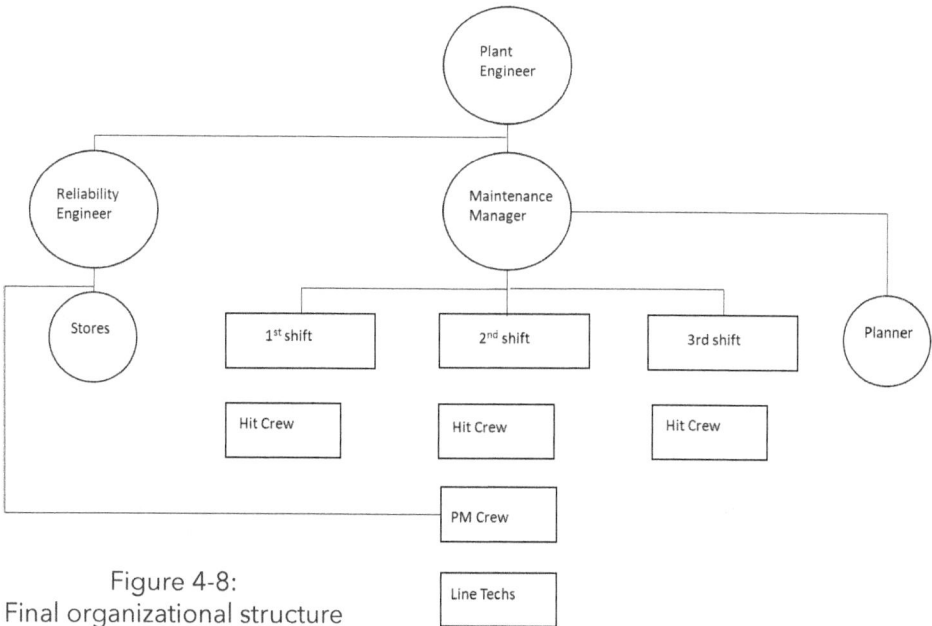

Figure 4-8:
Final organizational structure

We further agreed that if there was a breakdown, the lead man from the hit crew would confer with the lead man from the PM crew to determine if anything could be done to prevent another failure, or a failure on another piece of equipment.

We introduced a work order system (written) and, at the end of my time there, computers.

Lessons Learned

This really is the takeaway from this section on organizational structure. Much has been written on centralized maintenance, decentralized maintenance, value-stream centric, and even hybrid systems. What it really comes down to is, how are you going to ensure that your PMs get done without pulling technicians for emergency work? Also, how are you going to guarantee that the technicians who are working on a planned and scheduled job are not rerouted to another more important and immediate task?

It may be that the United States Marine Corps has had it right all this time. "It is essential that our philosophy of command support the way we fight. First and foremost, in order to generate the tempo of operations we desire and to best cope with the uncertainty, disorder, and fluidity of combat, command must be decentralized." (Warfighting, p. 79)

I had an unusual advantage in that I had thirty-four technicians. Most organizations don't have that many maintenance people. I also had an incredibly supportive boss and a very knowledgeable and patient consultant working with me. I forgot to mention that this was a union facility. I did not have unwavering support from the thirty-four people. I learned what a grievance was. There was a lot of work between me and Human Resources to develop the roles and responsibilities, and everyone in the department was allowed to bid on a job if they wanted it. This process took some time. But it stuck.

Earlier you recorded your staffing levels. Jack said to first carve out a PM crew that would do nothing but preventive maintenance. Name two people you would like to have on a PM crew:

List two people you would like to have on a hit crew (Do It Now squad):

_____

_____

Next, in the space provided in Figure 4-9, sketch out an organizational structure that would allow you to perform preventive maintenance, corrective maintenance, and emergency maintenance at the same time.

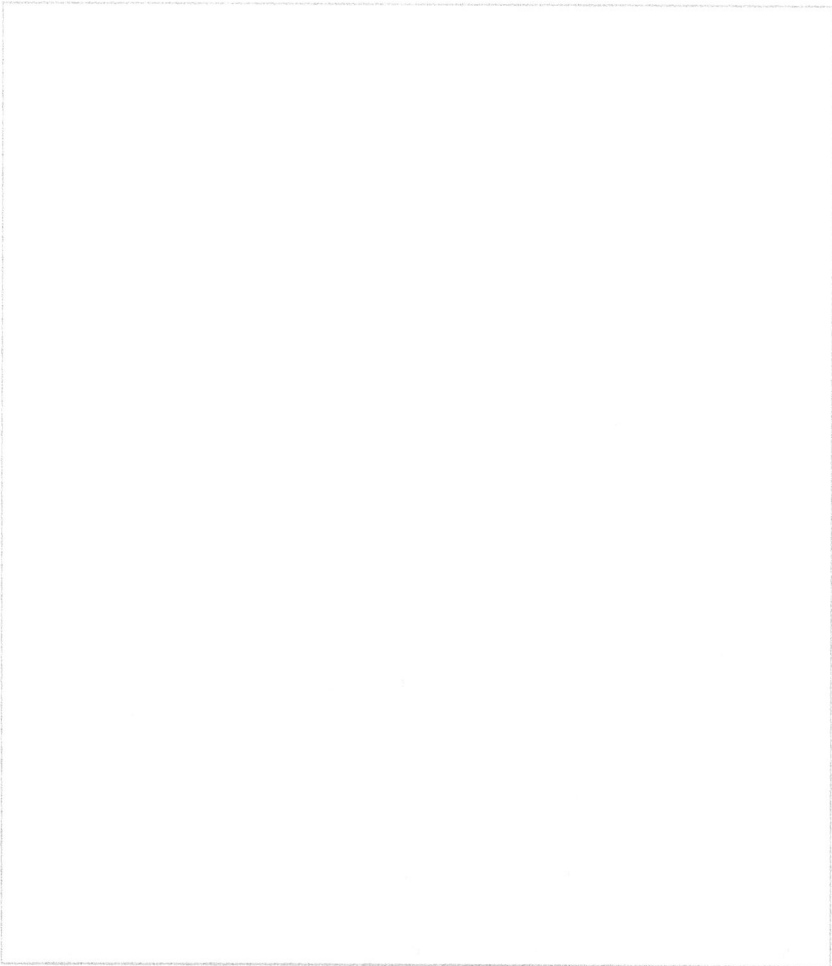

Figure 4-9: Draft organizational structure

Does it look good on paper? This is your winning organization-al structure!

Reliability KPIs

We are going to round out Chapter 4 with a most unusual discussion on Key Performance Indicators, specifically reliability KPIs. Recall in an earlier section we listed some principle elements of reliability:

- Design, build, and install an asset with the reliability of that asset in mind

- Replace components with like components

- Modifications should increase reliability

- Keep the asset clean, lubricated, and all fasteners properly tightened

- Work to increase the MTBF

There are many metrics and KPIs that can be used as leading or lagging indicators. This unique list of principal elements opens a door for a different set of measures that might not be as common as they should perhaps be.

If we know that designing, building, and installing an asset with reliability in mind is important, how exactly would we measure that? Do we have a standard of what is acceptable and tolerable for us and to us? Do we audit and hold the OEM to those standards? We should. We should also work to have a legitimate Bill of Materials on at least 98% of our assets.

Replacing components with *like* components is a surefire way to guard the inherent reliability of an asset. A pragmatic and responsible Management Of Change (MOC) process can ensure this actually happens. That process has to be audited and kept honest.

Modifications should increase reliability. As we've already imagined, OEE is an ideal candidate for KPI selection in this effort.

Keeping the asset clean, lubricated, and all fasteners properly tightened (think TLC) are all activities that can be routinely identified, standardized, and audited. The measures we use for these actions become the metrics.

Working to increase the Mean Time Between Failure can be measured by a simple calculation of the time span between failures. This would tap into the repair quality, either conducted in-house or contracted out. An effective and objectively-based training program would be a telltale sign of a promising MTBF effort.

To close out this chapter, list all the maintenance and reliability KPIs you currently track and report on in Table 4-8. In that table, also list how that KPI ties directly to reliability. You may be surprised that many maintenance KPIs have nothing to do with reliability.

Table 4-8: KPIs and their reliability connection

KPI	Reliability Connection

Chapter Summary

Equipment reliability often remains an elusive prize. It is very likely that many reading this book are predestined to fall short of that which they seek. As we've come to understand, an asset has to be reliable to begin with to be reliable. That might sound like an odd statement but considering the compromises made in the design and build phase of equipment, and the hasty install, it's a wonder our capital equipment runs as long as it does.

In this chapter we discussed the reliability ideas of:

- Equipment design excellence
- Inherent reliability
- OEE
- Maintenance and reliability strategy

- Cost benefit analysis

- Organizational structure

- Reliability KPIs

From our study and discussion, the following would be helpful to add to our developing strategy.

Our developing philosophy:

Our continued passion for reliability going forward has to be rooted in the past. We know that an asset has to be designed, built, and installed in a manner that provides the greatest level of inherent reliability.

As this is the case, we will contribute manpower and knowledge to those making equipment decisions to offer the best technical advice on *D—design*, for maintainability and reliability, as well as *B—build*, with robust components that are already proving

Machine:		Date:		OEE %	
Product:		Shift:			
AVAILABILITY					
1. Gross Available Time:	If 8-hour shift=480 min., if 12-hour shift=720 min				Minutes
2. Planned Downtime:	Time for meetings, cleaning, breaks, PM, and other scheduled events				Minutes
3. Run Time:	Gross Available Time – Planned Downtime	#1 - #2 =			Minutes
4. Downtime Losses:	a. Breakdown minutes _____ b. Changeover minutes _____ c. Minor Stoppages _____	a + b + c =			Minutes
5. Operating Time:	Run Time – Downtime Losses	#3 - #4 =			Minutes
6. AVAILABILITY:	Operating Time ÷ Run Time x 100	#5 ÷ #3 x 100 =			%
PERFORMANCE					
7. Total output during operating time:	Parts, yards, units run per shift: good & bad				Total Units
8. Theoretical Cycle Time (ideal):	Minutes per unit				Min./ Units
9. PERFORMANCE:	[(Theoretical Cycle Time x Total output) ? Operation Time]	[(#8 x #7) ÷ #5] x 100 =			%
QUALITY					
10. Rejects during operating time:	(machine related defects)				
11. QUALITY	(Total output - Rejects /Total Output. X100)	[(#7- #10)÷ #7] x 100 =			%
OEE					
12. OVERALL EQUIPMENT EFFECTIVENESS	(Availability x Performance x Quality x100)				%

Figure 4-4: (reprinted) Blank OEE form

their worth in our environment, and *I—install*, to ensure that our equipment operates in a space that is conducive to a long and reliable life.

In addition to other metrics and Key Performance Indicators, we will employ classic Overall Equipment Effectiveness to measure the combined value of Availability, Performance, and Quality both before and after an asset modification. We will use a form similar to this one to provide the measures:

Our maintainability and reliability strategies will work to reduce *Mean Time To Repair* and *Mean Time Before Failures*. We will win with an approach that provides the highest level of maintenance and technical guidance.

We will partner with other agencies to ensure that our advice and guidance is founded in good fiscal practice, always providing a reasonable cost with a benefit that enhances, improves or guarantees asset reliability.

Along with ensuring that we build equipment that is reliable at the start, and supporting the asset through its life cycle, we will create a maintenance organization that can simultaneously address preventive maintenance, corrective maintenance, and emergency maintenance.

To ensure that we are delivering on the reliability promises, we intend to develop, measure, and report on reliability specific KPIs and suggest measures that other departments can take to ensure their staff members are also supporting asset reliability.

FIVE

Leading and Organizing
for Reliability

All that we've studied and shared together in this workbook, up to this point, will be for naught if we cannot, or do not, lead our organization toward a higher level of maintainability and reliability. We explored the probability in the opening sections of this text that people do not necessarily object to change, they just simply don't want to be surprised by it, or more directly, be the victim of it.

For all the books and lectures that exist on the subject of leadership, it is interesting and a little troubling to note that leadership is still not a subject that is taught, front and center, in technical disciplines in most colleges or universities.

Business colleges have started to realize that they are behind the curve when it comes to the trends of the 21st century and now have popular courses on entrepreneurial doctrine and other offerings to capture the spirit of the younger generation. In a similar fashion, there needs to be a recognition among those involved in engineering, vocational-technology, and other such career paths that many of our young technically trained professionals are growing up to become the leaders of these newer, tech-savvy enterprises. What is troubling is that leadership is not a core subject in the engineering and technology curriculum. We need to change that.

Leading Vs. Managing

As corporations and businesses launch into the 21st century, with compounding pressures of global competition, it is important that the leadership and management principles of those companies alter their ratio of management to leadership to keep their businesses alive and viable. It is an entirely new world. Unfortunately, today's technical professionals might be ill-prepared for the duties of the day based on traditional education.

What is remarkable is the extensive level of coursework and corporate training available and dedicated to building hard managerial skills, but a comparative dearth of opportunity to find, build upon, and eventually transform those engaged in a technical field into leaders.

Recall the last corporate or company mandated training that you attended. Don't consider the usual CPR and HAZMAT training, but rather the advanced, professional training. What was that course, and was it management or leadership leaning?

Management/Leadership *(circle one)*

Are Leaders Born or Created?

Historically, companies have managed to the bottom line. Likewise, the evolution of many successful 'tech' companies came about through the charm and champion-like leadership style of charismatic entrepreneurs (e.g. Jobs, Gates, and Ellison).

Success on the heels of hard-charging and *devil may care* effervescence of these new-age leaders, compared to the quid pro quo drumbeat of managing to the bottom line, reveals a stark contrast in the differing talents as explained by Rear Admiral Grace Murray Hooper, "You manage things, you lead people." Nobody likes to be managed.

We Don't Seem to Hone and Sharpen Our Leaders

What has become clear in this century is that a successful outcome is more dependent on effective leadership than efficient management, but are we developing leaders today?

Identifying this shortfall, noted management contributor John Humphreys provided this foretelling, "Of greater concern, though, is the fact that some organizational decision makers simply have not acquired and/or developed the conceptual skills needed to operate as leaders in a world filled with ambiguity." ("Developing the Big Picture," *MIT Sloan Management Review*, p. 96)

In short, those in grooming for leadership positions are not actually being groomed to be leaders.

Humphreys' comments lend validation to the general thesis that leadership plays a greater role in success than does management. This idea might draw robust debate, but the state of today's fast and fluid environment cannot be simply managed. We are in the midst of a reliability revolution that must be led, not managed.

Who would you consider to be the leader of your organization; not the corporate CEO, but a more local person, someone with influence over your daily life?

Arent Managers and Leaders the Same?

The rub lies in the contrast between the classical differences of management and leadership. In the truest sense of action in the field, the lines begin to gray and the differences don't appear as stark near the border. In a combat theater, soldiers want to be led into battle by a leader, but the supply sergeant had better be a darn good manager. Both are necessary to win the war.

James Collins gives us a good starting point to establish our foundation for extolling the virtues of leadership and drawing a distinction between it and the very necessary skill of management. The effective leader, he tells us, "catalyzes commitment to and vigorous pursuit of a clear and compelling vision, stimulating higher performance standards." (*Good to Great*, p. 20) This is reminiscent of a field general. And as useful, the competent manager "organizes people and resources toward the effective and efficient pursuit of predetermined objectives." (p. 20) This can be likened to a supply sergeant.

Management—a Definition

Management, as a discipline, is a challenge for most. Fortunately, much has been written on what it takes to be a good manager. That in itself is a bit of a misnomer, given that merely reading a book on how to be a good manager does not make one a *good* manager.

Perhaps the most concise description of management is offered by W. Dale Compton, "While the tasks that management must perform are many and are often dispersed among many different individuals, the principal objective of management is to ensure that the organization is fulfilling its goals and objectives." (*Engineering Management: Creating and Managing World-Class Operations*, p.107)

There are three significant management responsibilities:

1. Strategy development

2. Employee relations

3. Financial oversight

Compton suggests another managerial issue, which is the need to continually train company personnel and press them to participate. He suggests that companies offer to reimburse employees who successfully accomplish educational goals. He also believes that employees, especially technically trained employees, will lose their edge after several years; therefore, their enrollment and involvement in continued education is critical.

In the space provided, briefly list the management and leadership leaning programs and training opportunities you and your associates are exposed to in order to strengthen abilities as a leader or a manager:

_____

_____

_____

_____

Fundamentally, and without exception, management is a critical function to all corporations, regardless of size. Every enterprise has a management function. Establishing an overall direction for the company with a defined strategy is essential to ensuring that the company is doing what it set out to accomplish. For the purpose of survival, managers need to keep the balance sheet in front at all times. The growth and development of the company personnel will ensure longevity and continued success for a thriving organization.

Draw the distinction between the management arts of strategy formulation, human resource elements, and fiscal responsibility, against the "indispensable" qualities of a leader, which are outlined in J.C. Maxwell's *The 21 Indispensable Qualities of a Leader.* (pp. V-VI):

1. Character: Be a piece of the rock

2. Charisma: The first impression can seal the deal

3. Commitment: It separates doers from dreamers

4. Communication: Without it you travel alone

5. Competence: If you build it, they will come

6. Courage: One person with courage is a majority

7. Discernment: Put an end to unsolved mysteries

8. Focus: The sharper it is, the sharper you are

9. Generosity: Your candle loses nothing when it lights another

10. Initiative: You won't leave home without it

11. Listening: To connect with their hearts, use your ears

12. Passion: Take this life and love it

13. Positive Attitude: If you believe you can, you can

14. Problem solving: You can't let your problems be a problem

15. Relationships: If you get along, they'll go along

16. Responsibility: If you won't carry the ball, you can't lead the team

17. Security: Competence never compensates for insecurity

18. Self-discipline: The first person you lead is you

19. Servanthood: To get ahead, put others first

20. Teachability: To keep leading, keep learning

21. Vision: You can seize only what you can see

A Quick Exercise

Consider this list of qualities. I'd like you to highlight those traits that you feel you possess and exhibit regularly. Really give this some thought. If you can, imagine a recent incident in which you were publicly or privately showing these characteristics.

Dr. Maxwell's guide is a twenty-one step process of sorts to help people become more effective leaders. His approach is unique in that he instructs the reader to not read his book, at least in its entirety, at first; rather, it should be read one chapter at a time. Each chapter teaches a different quality, and he encourages readers to take up to two weeks to digest and understand the teaching.

If leadership is intrinsic to our evolutionary make up, how is it fed, nurtured, and matured? Very little instruction, certainly elementary instruction, is geared towards allowing children to develop their budding leadership qualities. Little changes through secondary and post high-school life, unless a student seeks out these opportunities through sports, student council, in-school clubs, extra-curricular activities, or part-time employment. It might be argued that the very act of seeking these extra-curricular activities is a sign of leadership.

360° Leadership

Larry Bossidy, co-author of *Execution: the Discipline of Getting Things Done*, takes a different approach in examining the art of leadership. His article not only reveals what a leader might expect from you (each of us), but counters it with what "my direct reports can expect from me." Clarity of direction is his first deliverable. "If I'm the leader, it's my job to communicate clearly where the business is going, why, and what the benefits will be if we accomplish what we set out to achieve." (p. 64) Bossidy continues with the following, "setting goals and objectives; giving feedback; be decisive, accessible; and demonstrate honesty." (p. 64)

Listing the needs that a leader has from his or her people is the major thesis of Bossidy's work, and one that makes an interesting discursion to the discussion of leadership. What your leader expects from you is tangential to the qualities of leadership insomuch as what is expected from your boss might very well be the very qualities you must exhibit when you are the boss.

Get involved by knowing when to delegate and when to take the reins of control. Bossidy writes from the point of view of a CEO, so his subordinates are themselves executives in the organization. Generating ideas is another demand he places on the people in his organization. Lack of ideas, he says, is a major frustration to corporate America. His people must also be willing to collaborate, share credit, and work as a team for the greater good. Be a leader, or more specifically, be willing to lead initiatives. Bossidy's executive team members must themselves be developers of the next level of leaders. His discussion on this topic ends with a quick touch on action phrases: stay current; anticipate; drive your own growth; and, be a player for all seasons. (pp. 60-62)

Imagine for a moment that you are the leader of the maintenance and/or reliability effort at your location. In fact, you may very well be the leader. If your team were to provide you with this 360° feedback, what would you like it to say? Record those thoughts here:

_____

_____

_____

_____

Leadership, like management, is intangible. It is, in itself, an essence that requires a practice that is accomplished on faith. By faithful demonstration of leadership efforts, people will be encouraged to be 'led.'

The very antithesis of accepting the status quo is leadership.

Leadership in the 21ˢᵗ Century

Who are these leaders of today, and why do we need them? Seemingly, the ability to learn some classic leadership traits and a little practice through on-the-job training would get business professionals off to a good start. However, there exists a fine line between the need to manage and the need to lead.

There is a hunger for leadership in business. Managers have to lead, although it's becoming more and more obvious that those great managers woefully lack the necessary leadership tact to move forward. By no fault of their own, we are growing managers at an implausible rate.

Steve Coats (www.i-lead.com/articles/article002.html) paints a very concerning picture, but one that resonates through much of industry today.

> "Companies are investing staggering sums of money in quality initiatives, process and systems re-engineering, and customer driven reorganization. Sadly, a large proportion of these strategic overhauls have not lived up to their promises. The reason is they desperately lack the visionary leadership required to strengthen people's belief in a new future, and their confidence that the new strategies will take them there. With no leadership, people get exclusively focused (and easily swallowed) by management directives of streamlining the "hows" [sic] of the business. The meaning of "why" the work is important gets lost. With down-sizing a closely associated outcome of efforts like these, people quickly become cynical about the future and turn away. When this happens, disappointing results are inevitable.
>
> Many companies have decided that the best way to manage the challenges posed by these "re-everything" efforts is to throw teams at them. The value of teamwork in today's world is beyond question. It is a fact that more and better work can be accomplished through teams. The world is just too complex for even the best to fly solo and longer, thus high-performing teams doing real work offer tremendous advantages. This is especially true if the senior officers can find ways to work as a team, because they too will be able to get more done, as well as visibly model the value of teams for everyone else.
>
> But, for any team to be successful there must first be effective leadership. Leaders provide vision, competence and heart, enabling and inspiring members to want to achieve astonishing performance results."

There is in fact a growing need for leadership. A need that has been identified and exemplified through example, but how do leaders develop?

Where Do We Find Leaders Today?

An eminent source of great leadership training in the United States has conventionally been the military. Incontrovertibly, General MacArthur's farewell address to the cadets of West Point, relaying his sentiment that, "The long gray line has never failed us," was reminiscent of a school whose tradition was one of great leadership, not great management. It might be argued that the greatest results have not generally manifested themselves as a result of superior management.

Even today's young people have great opportunities for leadership development through school activities well before college instruction. Some schools have hallowed histories of producing some of the finest leaders in the nation. "Leavenworth [Kansas] High School boasts at least 33 generals who have passed through its doors as students and graduates. School officials think that's a public-school record. They credit neighboring Fort Leavenworth and the school's distinguished JROTC program with laying the groundwork for disciplined leaders." (D. Bormann, "Turning Out Generals," *The Kansas City Star*, p. B1)

The maturation of leadership abilities might begin in a classroom, but they most certainly are recognized when given an opportunity to be exercised. If not, however, we must give these budding leadership opportunities a chance for life in the work place.

While character and knowledge are necessary, by themselves they are not enough. Leaders cannot be effective until they apply what they know. What leaders say or do is directly related to the influence they have on others and what is done. As with knowledge, leaders will learn more about leadership as they serve in different positions.

If you were to plot a career trajectory for yourself, in your current organization, over the next twenty years, what would it look like? List it here:

Clearly, any organization's obligation to teach and grow leaders is essential to success. How important is leadership over management to your organization?

Maybe We Need Both Management and Leadership

The crux of the argument is the head-to-head comparison of management skills versus leadership traits. A reasonable assertion could be made that the difference between the two is infinitesimal and different situations would bring into play the different principles. A counter to that argument is simple acceptance of the relevant fact that if management or leadership abilities are not taught and developed, they cannot be brought to bear. Each has its time, and there is an art to recognizing the need to evolve.

> "Leadership and management are both important, but they seek to do different things. Every organization structures itself to accomplish its goals in a way that is in tune with or responsive to its environment. Once the efficiency of the organization is established, people go about simply maintaining the system, assuming that the environment will stay the same. Management is the main focus because it keeps the organization going well with little change. But the thing is: the environment for any organization is always changing. There are always shifts in consumer tastes, social attitudes, society's culture, technology, historic events, and so on. The world is not static as we assume. Organizations tend not to spot these changes quickly, often because of a "management orientation" which is focused more on "looking in" instead of "looking out." Over time, the organization can become less and less in tune with or responsive to its environment, creating more and more management problems. Times like this require organizations to think more in terms of leadership. Leaders begin to ask questions like, "What is really going on here?" "How do we become relevant again?" "How do we fulfill our goals in these new times?" "What will prompt people to think that what we do is meaningful?" Leaders seek to bring their organization more in line with the realities of their environment, which often necessitates changing the very structure, resources and relationships of their organization which they have worked so long and so hard to manage. And yet, as they do, leaders can bring renewed vitality to their people."

> (www.telusplanet.net/public/pdcoutts/leadership/LdrVsMngt. html)

An irrefutable need exists in an organization for both management and leadership. The two ideologies are not diametrically opposed, and with little exception, the person responsible for *management* is also the person responsible for *leadership*.

> "Leadership is the WHAT and WHY we do things while management deals with the WHO, WHEN, WHERE. Leadership establishes the direction of the organization and management handles the logistics needed to get there. Leadership establishes the vision; management provides the hands. Leadership inspires and cheerleads; management coaches. Leadership inspires others to go beyond what they thought they could do while management gets others to do what they need to do. Leaders blaze the trail while managers keep the trails open and supply lines strong. Leaders are able to raise [sic] above the fray to see the entire campaign while managers concentrate on the battles at hand. Managers concentrate on this day, week, or month while leaders are already living the next one, two, or five years ahead. Managers get things done; leaders plan what things need to get done. Managers climb the ladders; leaders make sure the ladders are against the right walls. I do believe that certain jobs and responsibilities belong to leadership and others to management. I also believe, however, that good leaders keep communication lines with management open so that their input can be considered as leadership considers and plans its next action.
>
> These beliefs do not negate the fact that one individual may fill either role in different contexts. One may be a leader in one campaign while filling manager shoes in another. In fact, unless we sign the paychecks, it's likely that we play both roles within our organization. But while one is in the "leader" role, there are different responsibilities, tasks, and skills that are utilized and brought to bear than when one is acting as a manager."

> (http://hwebbjr.typepad.com/openloops/2005/03/my_two_cents_le.html)

The historic bases of teaching and growth in an organization is geared more toward managing and maintaining a status quo with little opportunity for leadership development.

This gap is even more evident in technical fields where students and young professionals are invested in learning the rapid degree of core information that comes at them, in an effort to keep pace with the changing times. Often, if successful in their primary roles, these same individuals are elevated to levels of management, having never been given a single rudimentary exercise in leadership. Hi-tech industries can be worse.

> "Never before in history have there been such profound knowledge gaps between managers and the front line employee who create business value. In earlier times, supervisors often rose from the ranks of workers and therefore were experts in the activities being supervised. But even when that is true today— when a software developer becomes a development manager, or a genetics researcher becomes a research manager—the specialized field in which value creation occurs keeps moving forward." (R. Austin and R. Nolan, "Bridging the Gap Between Stewards and Creators," *MIT SLOAN Management Review*, p. 29)

A Gap Analysis

Recognizing a compelling gap between the effectiveness of management and leadership, Joe Reynolds wrote, "A 1991 Leadership Studies International survey of 700 top corporate executives found 60 percent were dissatisfied with their personal leadership efforts. Could these executives feel that their success was more a result of good management than leadership, and that stronger leadership skills would have produced even greater success? I would bet on it!" (*Out Front Leadership—Discovering, Developing, & Delivering Your Potential*, p. 29).

"[The] leader establishes and conveys the values and vision of the organization, and develops a master strategy to achieve the goals of that vision. The leader then empowers capable managers to develop and implement strategies for their departments that support the master strategy." (p. 30)

Leadership and management are complementary; they co-exist for the benefit of the organization. Reynolds' last entry in the previous paragraph indicates that the leader and the manager might be different people, or at the very least that the approach might be different given the differing circumstances.

"Decidedly, the world today is a different place than last year, or the year before, or even a decade before. One year's time seems to bring on several years' worth of change—that's how fast the process of living has progressed. What worked before will not necessarily work tomorrow. There is a new world order.

In today's new economy, the old ways of management no longer work and will never work again. The magnitude and pressure of environmental, competitive, and global market change we are experiencing is unprecedented. It's a very interesting and exciting world, but it's also volatile and chaotic. You cannot address these new challenges with more of the same management solutions—successful change requires leadership."

(www.1000ventures.com/business_guide/crosscuttings/leadership_vs_mgmt.html)

The head-to-head comparison of management vs. leadership strengthens the position that both qualities are needed for success. Given that we get more education in management than in leadership, the impact of this shortcoming has to be measured.

Although both management and leadership are required and desperately needed in the fields of maintenance and reliability, we need to lessen the judgmental tendency we have to do damage to ourselves and instead promote the notion that ours is a professional discipline that needs leaders.

Failure to recognize this can stymie our improvement efforts, or flat out kill them. Mary Walton said as much in her work. "Most managers, when they view their group, are supervising, judging, and ranking the performance of the individual workers. But a leader judges his own performance when he observes his group. In his mind he is determining what he has to emphasize or de-emphasize, what action he has to take to foster improvement." (*The Deming Management Method*, p. 176)

Sometimes brevity is the best course. The quintessential high-water mark of a leader, Jack Welch, has this simple message. Leaders, he says, "Inspire with clear vision of how things can be done better." (Robert Slater, *Jack Welch and the G.E. Way*, p. 29).

Staffing

One of the best ways for leaders to earn their bona-fides is to convince those ultimately in charge that the maintenance and reliability divisions need to be properly, completely, and professionally staffed. This human resource is without exception the most critical to high performance.

Often heard in the effort to loosen the corporate purse strings is, "You don't cost cut your way to world class." There must, however, be some throttling or metering method to thoughtfully create an effective maintenance organization.

Some companies have failed to see that maintenance and reliability are of core value importance to the operation. It's true that a plant can run without maintenance for some time. But for how long?

Whether a company chooses to run a lean organization, or is staffed to the hilt, we maintenance and reliability professionals are mandated to show value for money.

In an earlier budget discussion, we introduced the common terms of direct and indirect labor. For most organizations the direct labor is inclusive of the hourly workforce. It is their work that contributes directly to the making of the 'thing' or the providing of the service. Indirect labor is usually the salaried staff, and there is no real direct line connecting their work (or existence) to creating a product or service.

It would undoubtedly make sense that the majority of any workforce be direct labor, and therefore link to the creation of the product or service.

Quality vs. Quantity

As it provides insight into the proper staffing levels to achieve, we might be doing more harm than good by focusing on the correct count of direct and indirect labor to have in our maintenance organization. We should, I'd suggest, concentrate on our vision and what roles will be necessary, not only to get there, but to have available when we are there.

An Example

Years ago I was conducting an assessment of an East Coast manufacturing facility and I had a week-long series of meetings with each of the twenty-six maintenance technicians. During those talks, each person told me that the organization needed to hire more maintenance people. When I inquired as to why, the response was exactly the same from each, "So we can handle more breakdowns."

I asked each group of hourly technicians if they thought a consultant could get anywhere with plant leadership in a request to hire more maintenance people so they could handle more breakdowns.

It seems kind of silly when you phrase it like that.

A Better Way Forward

Rather, I offered this as an example: why don't we create a preventive maintenance team and make an argument that we need to back-fill three maintenance technicians. The PM team will be out in front doing work to reduce breakdowns.

Can you see how this approach might lead to additional maintenance people *and* a preventive maintenance team *and* a plan to reduce emergency work? The argument might not hold sway with upper management, but it makes for more of a compelling case.

The Numbers

There are some reports that indicate that a maintenance staff should make up about 18% of the hourly workforce in a facility. This is an applicable value regardless of the industry. There are other standards that suggest specific staffing for facilities maintenance organizations based on the square footage of the facility or facilities. These are all valid numbers, and I've used them myself.

Where our intent begins to cloud is the justification to have such a well-staffed organization. Consider that when a maintenance organization performs as it should, absolutely nothing happens. Of course what is meant by this is that a well-staffed, well-functioning maintenance organization seems to be docile and smooth running. From the outside, it might appear to the casual observer to be overstaffed.

Imagine justifying an ideal level of maintenance technicians and staff and keeping that level when everything runs as it should. Has it ever happened that a financial analyst has recommended cutting the predictive maintenance contractors because nothing is breaking?

I was teaching a class in North Carolina when, during a break, one of the students came up to me with his phone and told me that his boss wanted to talk to me. "Hello," I said, "This is John, can I help you?"

The voice on the other end was not friendly. "Where do you get off telling my maintenance supervisor that we need to hire more maintenance people?"

Of course, no such conversation ever happened. Someone had asked about that 18% standard for maintenance staffing and I showed the class where it was recorded. I'm very certain that the standard included Equivalent Man-Hours for the thousands of hours operators spend with 'look, listen, and feel' preventive maintenance tasks.

As it relates to standards, statistics, and world-class numbers, we must remember that those values are meant to cover a large swath of industries and company sizes. A three person maintenance organization is most likely not going to form a PM crew, a hit crew, and a corrective maintenance crew.

Where we do have an advantage and can gain some ground is in establishing a maintenance staff, on paper, and making a direct link between the formation and performance of that team with equipment reliability. When our staffing ideas are on paper, we can run a multitude of 'what-if' scenarios and demonstrate to those in maintenance and those served by maintenance how the alignment will be aimed at equipment reliability and ensuring that the facility runs as needed.

We are going to meld some earlier points together here to really hash out an effective and efficient maintenance staff. Keep in mind that for every role that is defined, it isn't completely necessary that you have a person that performs that role and only that role. In many small maintenance organizations people may wear a lot of different hats.

You don't have to staff each position with a dedicated person.

It should be asked, though, how many jobs a person can do at once, 100% correctly.

Maybe We Should Determine Our Structure Before Our Staff Levels

Recall our earlier review of roles and responsibilities in Chapter 3. We spoke specifically about the many roles that a maintenance planner/scheduler might have. Within each of those individual roles are detailed responsibilities. Combine that thought with a more recent discussion in Chapter 4 regarding an effective maintenance organization such as the one I established at my first civilian job. We are going to use that organization simply as an example to help start our conversation on staffing.

Figure 4-8 is reprinted for your convenience.

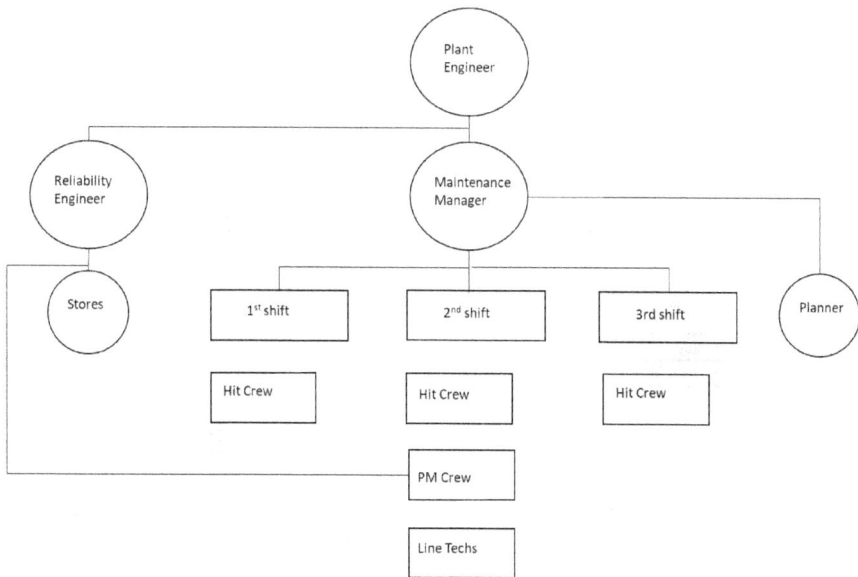

Figure 4-8 (reprinted): Final organizational structure

Again, just as an example, we are going to explore some fundamental elements of this particular organizational structure and establish, arguably, some very good elements to consider in your structure with your staffing vision.

The staffing for the structure shown in Figure 4-8 will be filled out with mechanics, electricians, supervisors, and store clerks, but primarily, the positions required are:

- Plant Engineer
- Reliability Engineer
- Maintenance Manager
- Storeroom Supervisor
- Planner/Scheduler

- Shift Supervisors
- Hit crew lead men
- PM crew lead man
- Line Mechanics

In these positions, each person may have roles as shown in the next series of tables; starting with the Plant Engineer in Table 5-1 and continuing to Table 5-9, the Line Mechanic.

Table 5-1: Plant Engineer

Plant Engineer
Department director
Mentor
Organizational structure developer
Budgeter
Stock determiner
Report generator
Chief documenter
Facilitator
Project manager
Supervisor

Table 5-2: Reliability Engineer

Reliability Engineer
Mentor
Assistant organizational structure developer
Assistant budgeter
Assistant stock determiner
Assistant report generator
Documenter
Facilitator
Assistant project manager
Supervisor

Table 5-3: Maintenance Manager

Maintenance Manager
Mentor
Assistant organizational structure developer
Assistant budgeter
Assistant stock determiner
Assistant report generator
Documenter
Facilitator
Assistant project manager
Supervisor

Table 5-4: Shift Supervisor

Shift Supervisor
Mentor
Assistant organizational structure developer
Assistant budgeter
Assistant stock determiner
Assistant report generator
Documenter
Facilitator
Assistant project manager
Supervisor
Reporter of crew availability

Table 5-5: Planner/Scheduler

Planner/Scheduler
Planner
Scheduler
Facilitator
Maintain BOMs
CMMS gatekeeper
PM/PdM overseer
Picklist generator
Data analyzer
Report generator
KPI/Metric generator

Table 5-6: Storeroom Supervisor

Storeroom Supervisor
Mentor
Assistant organizational structure developer
Assistant budgeter
Assistant stock determiner
Assistant report generator
Documenter
Facilitator
Assistant project manager
Supervisor
Stores Stock Committee Chairman

Table 5-7: Hit Crew Lead Man

Hit Crew Lead Man
Mentor
Assistant organizational structure developer
Assistant stock determiner
Report generator
KPI/Metric generator
Facilitator
Documenter
Executor (of the maintenance plans)
Reporter of crew availability
Supervisor

Table 5-8: PM Crew Lead Man

PM Crew Lead Man
Mentor
Assistant organizational structure developer
Assistant stock determiner
Report generator
KPI/Metric generator
Facilitator
Documenter
Executor (of the maintenance plans)
Reporter of crew availability
Supervisor
Developer of PM strategies

Table 5-9: Line Mechanic

Line Mechanic
Master of the line
Assistant stock determiner
Report generator
KPI/Metric generator
Facilitator
Troubleshooter
Repair person

Once you have laid out the roles that need to be covered in your reliability effort, a list of corresponding responsibilities needs to be created. A combination of roles and responsibilities, along with facility size, asset count, and maintenance backlog work will provide the calculus required to fully and completely determine the proper staffing levels.

Organizational Capability—Skills Matrix

There is not a single facility around the United States, or in the international companies I've worked with that does not report a dearth of talent in the skilled trades. There has to be a great multitude of reasons for the state of our skills deficit, maybe even some good ones. However, the sum total of whatever is happening is that we are running out of maintenance people, and we don't seem to be growing any new ones.

As our factories reach for the 22nd century, our technical workforce is reaching levels not seen since the 19th century.

Technology Doesn't Make Everything Better

Give some thought to the expanding gap between what we currently know about technology in our facilities now, and what the project engineers are working on right this minute. We've already discussed the idea that we don't do enough to standardize parts, or even try to stick to a common design.

It's been said that by the time you install a new machine, the controls technology is already obsolete. Recovering precious metals from discarded circuit boards is a booming business.

The machines are taking over. If it's not already here, Armageddon is just around the corner.

How on Earth Did We Get Here?

I attended a conference recently where the President of Marshall Institute was delivering a presentation on this very subject. During his Q & A session, one of the attendees in the back made a very interesting observation. This was his firsthand accounting.

This fellow attendee began by telling the group that he was currently a high school guidance counselor. "Some of the counselors," he said, "use the technical trades as a threat. My colleagues will say, 'You better buckle down. You don't want to be a welder your whole life. Do you?'"

When did the honorable skilled trades go from highly praised and respected to just a step above reform school? My goodness, what are those counselors thinking? Worse, what are they doing?

Someone should have set them straight and reminded them that really good welders end up owning a welding company and hiring several other welders. They are entrepreneurs and job creators, valued members of the machinery that runs this world.

The United States has moved from a manufacturing-centric economy to a service centered one. The loss of these skills is moving us from a producer of the world's goods to a net consumer.

But Wait, There Is Good News

We know how to operate in this environment. How do I know that? Because we have come together through over two hundred workbook pages developing a strategy and a plan this entire time. We are going to nail this as well.

Embrace the Change

We know that technology is not going to be reverting to the era when our equipment was designed, built, and installed. No, not ever. Instead we are facing the IoT, the Internet of Things. If that doesn't sound like Skynet (*Terminator* movie reference), I don't know what does.

We know this is coming and we can be prepared for it.

For example, when we design a new machine, or make a modification on a piece of equipment, we need to consider that the technology we have in place might not be what we want for the new machine. For one thing, it most likely has limited after-market support.

Instead, and recall the organizational structure we worked on, we are going to have a Reliability Engineer who is directly linked to the storeroom and to the PM crew. We definitely have a horse in this race.

Play to Your Strengths

A lot of organizations are tapping their production crews to find that diamond in the rough. Every facility has *that* one person, or those people, who exhibit the highest interest in their equipment and are just a pleasure to work with.

List the first name of one of those great production associates:

_____

This operations-based, ad hoc work force is essentially a built in stable of resources for any maintenance department that is responsible for keeping its company's equipment running reliably.

There will always be the thinning supply line of ready and able mechanics and electricians to consider. Let's face it, in some ways we've done damage to ourselves by downsizing our own internal capabilities.

Many of you reading this text can remember a time when companies would have mechanics, electricians, controls folks, plumbers, pipefitters, carpenters, and painters. Today, we've boiled that down to just two or three catch-all trades. A person no longer has a three year apprenticeship in a single trade. They have a learning curve so steep that it looks like a mid-runway, military jet launch (straight up).

The line of skilled persons walking through our factory doors is thinning. More often than not, if there ever was an unusually large supply in the local area, that would mean that some company in the area just had a layoff.

There is, however, one group that is always growing: the production staff.

A Really Good Idea
I was the plant engineer responsible for building a new plant, many years ago. One of the ideas my plant manager had was to partner with a local temporary agency who would work as our associate vetting service. Employees would come on board and be in the usual probationary period, but the probation period was not with us, it was with the temporary agency.

My boss explained that this gave the employee six months to see if they liked us, and we had six months to do the same.

I work with a client who had such a good idea I wish that I had thought of it. They will hire folks that come in through the usual channels. If the new person is already a skilled worker, they take a test, and then, based on the openings, they come on board. Operators and other non-skilled associates in the company have to start in the storeroom. They stay in the storeroom until there is an opening available in maintenance for which they can apply.

This is such a good idea. It bolsters the storeroom staff and it leads to a motivated workforce. Not to mention that when these folks get into maintenance, they have a really good idea about the spare parts offerings, and how *not to* mess up the storeroom.

FJA

Regardless of the path our people take to get into the maintenance organization, once they are there we have to train them.

Most have heard of OJT, *On the Job Training*. Real OJT requires technical manuals, a training scheme, signatures, audits, and objective evidence that the learner has learned.

I've just listed some attributes to real, formal and successful OJT. Be concise, yet thorough. In the space provided describe your organization's OJT program (if you have one):

_____

_____

_____

_____

I contend that if you do not have the elements I've just described, you don't have an OJT program and instead you have an FJA program: Follow Joe Around.

"Hi Jerry, welcome aboard. Grab your tools and follow Joe around for two weeks."

Just to seal our fate, we'll move Jerry to the midnight shift after two weeks and install him as the maintenance expert on the midnight shift.

Cost/Benefit Relationship

Our attitudes about skills training are starting to turn a corner. They have to. We've come through the forest of cost cutting to reach a clearing of sober thought and lucid consideration regarding how we've put ourselves in a very tight spot.

Rafael Aguayo cautioned in his study of Deming that, "Training is often seen as an expense. It is a visible number that management can control and therefore subject to scrutiny. Training often has to be justified, but the benefits from training are often not visible." (*Dr. Deming: The American Who Taught the Japanese About Quality*, p. 168)

Even Deming himself said in his book *Out of the Crisis* that, "The greatest waste in America is failure to use the abilities of people." (p. 53)

Deming also instructed leadership to do more to enable their workforce and then to remove road blocks, saying that, "Money and time spent for training will be ineffective unless inhibitors to good work are removed." (p. 53)

There is an absolute certainty that our maintenance technicians of the future are going to require a higher level of learning, teaching and growth opportunities. Let's face it, our maintenance workforce today is smarter than it was in the past. I don't mean any disrespect; we've always had extremely talented, intelligent and capable skilled workers, but they've gotten smarter as we all have gotten smarter and more capable. Heck, the internet alone has given all of humanity a veneer level of knowledge on everything imaginable.

The skilled workers in today's factories are likely to have college degrees, skills certificates or vo-tech experience. Shop classes in high schools might be a thing of the past, but our direct labor is sharper than ever.

As such, it isn't beyond reason to expect our workforce to perform to a level commensurate with what they are paid. To make that connection, our skill-based training has to be competency based.

You Should Get What You Pay For

Our skill-based training process should result in a true skills mastery. Suzuki's edited work, *TPM in Process Industries*, informed us that, "*Skill* is the ability to do one's job, to apply knowledge and experience correctly and reflexively in all kinds of events over an extended period. Systematically accumulating training, experience, and information enables a person to exercise good judgement and act appropriately. The more swiftly a person can deal with an abnormality, the higher the skill level." (p. 262). And that, "Skill is the product of personal motivation and thorough training. The end result is mastery. To enable people to achieve mastery, companies must develop the most effective training methods." (p. 262)

The Training Gap Analysis

If we are going to work within our organizations to develop an exceptional skills training process, we have to first establish our starting point. It might sound overly simplistic, but we do this by asking our current workforce what they feel they need to know in order to perform their current jobs.

Give this a try. In the space below, see if you can list five individual abilities that a mechanical technician needs to know in order to be a mechanic at your location. I added a common one to help you get started: Remove and replace a gearbox.

_____

_____

_____

_____

That should have been simple enough. To establish a listing of what *skills* your workforce needs to possess to be a skilled worker at your location, simply ask them. You have to clearly communicate to them, however, that you are not asking them what they know how to do, but rather, what they need to know how to do.

For example, it might be that you have a mechanic that is proficient in AutoCAD. That most likely is not a required skill for a mechanic to have at your site.

After your associates are polled as to their thoughts on what is required to know at your location, then the maintenance and reliability leadership edits that compiled list to come up with what the *leadership* expects of its skilled workers. The leader's role in the training process will become more of a supporting, enabling, and coaching role, or as Walton says, "The job of a supervisor is not to tell people what to do or to punish them but to lead. Leading consists of helping people do a better job and of learning by objective methods who is in need of individual help." (*The Deming Management Method*, p. 35)

The abilities that make the final 'cut' should be listed on a spreadsheet that clearly defines what skills or abilities a tradesperson is required to have to be in the maintenance department. Don't forget to record some of the softer skills such as being reasonably proficient in Microsoft Word, Excel, signing in and out of the CMMS, and entering work order information.

What follows in Figure 5-1 is an example of what that spreadsheet might look like:

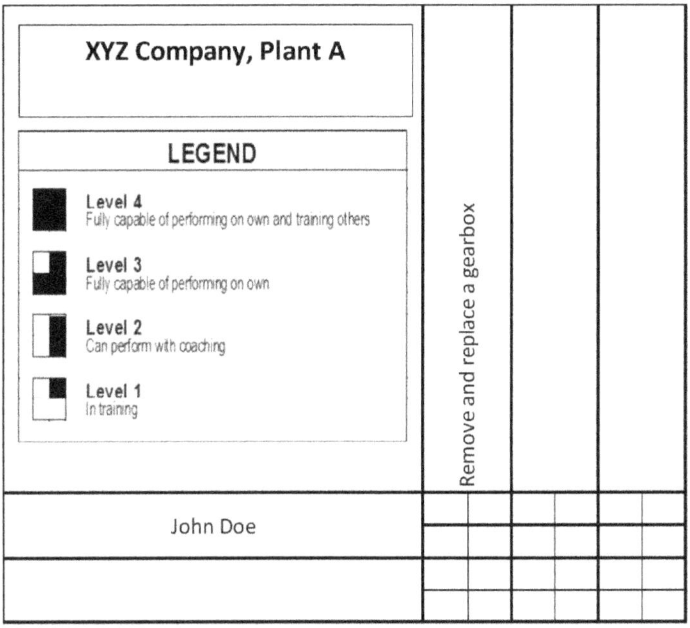

Figure 5-1: Skills Matrix

You may have noticed in Figure 5-1 that there is a legend in the upper left-hand corner. More on this in a minute. The job we, as the leaders of the maintenance workforce, are to do is to populate the columns and the rows of this spreadsheet.

The columns will contain all the abilities our folks are required to have, and the rows will be filled with the names of all the applicable associates.

Remembering Ron Moore's point in *Making Common Sense Common Practice*, "... when training is provided, expectations should be articulated about the consequences of the training on improved productivity, output, etc. Training for its own sake, while philosophically satisfying, may not be in the best interest of the business." (p. 409)

Our training will only net the results of increased output if the newly minted skill is exercised and used for the good of the organization. We do this by requiring a level of mastery. I would recommend there be a stepped-pay increase based on the demonstrated ability to perform a certain task at an advanced level.

The legend mentioned just a bit earlier is our indication of the level we rank our employees. In brief:

Level 1—In training

Level 2—Can perform with coaching

Level 3—Fully capable of performing on own

Level 4—Fully capable of performing on own and training
others

We don't want, nor do we need an entire workforce at level 4, but we do need some people at level 4 in order to train others. The majority of our population should be at levels 1-3.

Figure 5-2 is what a more populated skills matrix looks like:

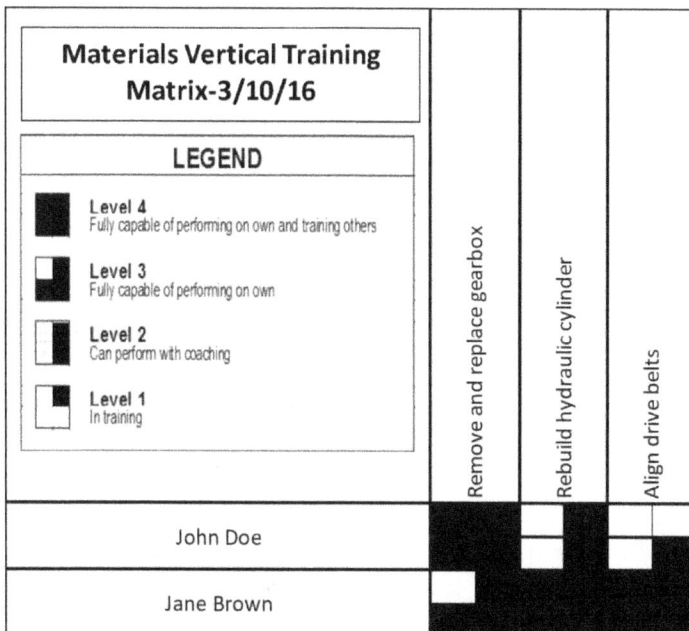

Figure 5-2: A more populated skills matrix

How can we populate the level of knowledge on our skills matrix spreadsheet, you may ask? The answer might seem unusual, but you ask everyone where they think they are. That does seem odd. It really is only applicable for the first time this matrix is assembled. After that, all rankings and level advancements are based on objective evidence of the skill. Think of it as a one-time amnesty on knowledge.

One Major Obstacle

We've been burned by training before. Creating an entire enterprise around growing a strong and capable workforce, only to have it disappear, is something we have all experienced.

Aguayo reminds us, "I don't think we fully appreciate the benefits of training. This is reflected in some of our common attitudes. Many firms like to hire from the outside, bringing in someone who has already been trained. That way they avoid the expense of training employees. There is a fear that if they train someone, that person will leave and apply the training elsewhere. In general, the major benefits of training are not well appreciated." (*Dr. Deming, The American Who Taught the Japanese About Quality*, p. 172)

What if we don't train them and they stay?

A Lesson From the Children

I want to close this section on training in general with a short story. My son is an Eagle Scout, and like all Eagle Scouts, he worked hard to make his way up the ranks. I am very proud of him, for that and other reasons. I'm proud of my daughter too, for her accomplishments in life.

But it's my son's Boy Scout experience that I want to share. When he moved from his Webelos to the Boy Scouts, his Webelos den worked its way around to visiting the local Boy Scout troops to give the young men an idea of the different troops they could join, if they had further interest in continuing with the scouts.

My son chose a troop for the sole reason that during his visit, they made a fire inside their scout hut!

As it turns out, this was a great troop to get involved with. This particular group of young men had an adventurous streak and were *all in* as it related to giving each scout a taste of everything you can experience. They had one unique element to their troop that I've always remembered and I tell this story often.

Inside their scout hut, one of the former scouts, as an Eagle project, had designed a large board, a sign really. Across the top of the sign was every single badge that a scout could earn. Down the left side of the board was a wooden plaque, containing the last name of each scout. The intersecting rows and columns indicated which scout had which badge.

Now, to attain Eagle Scout, a scout must earn several merit badges. There are ten in particular that are required. On the board, those ten were the first ten across the top. The rest followed.

In order for a scout to earn a merit badge, the scout wanting to learn the skill had to seek out a scout who already had successfully demonstrated the skill. The learner was required to seek out the learned. In turn, the scout who already possessed the badge was required to agree to be the mentor of the scout seeking the badge. The scouts, after all, are teaching leadership.

The point I want to make with this story is that the person requiring the skills and knowledge has a duty to seek out the skill and knowledge. This can't all be mandated by management. There is some onus on the part of the learner to seek out the knowledge.

Chapter Summary

This was an interesting chapter. It is reasonable to believe that not many of us have spent much time pondering the difference between leadership and management. It might be fair to think that the differences, no matter how slight or significant, have very little impact on our day to day lives in reliability. Ask yourself who would be more influential in demanding more resources, attention, and greater respect for equipment reliability in your organization: a good manager or a good leader?

It's important that we know the difference as we aspire to greater roles and more responsibility in our individual organizations. Both management and leadership are needed to move us deeper into the 21st century amidst technology that requires that we are constantly on our toes and aggressively seeking to maintain the highest level of maintenance on our assets.

Leaders or managers have to lead or manage someone, right? Our chapter continued with a discussion on staffing with a section regarding some of the more common roles that are found within maintenance and reliability organizations.

One area that is almost always at the root of our incomplete maintenance staff is the lack of skilled tradespeople available today. To combat this, we rounded out our chapter with a discussion of building a training matrix and identifying the gaps in our coverage. Closing the gap should be the focus of our skills training program.

Our developing philosophy:

To ensure that the proper level of reliability and maintenance management is available along with a balance of the necessary leadership in these disciplines, we intend to develop and execute a clear path of development curriculum that would include such courses, but not limited to:

_____

_____

_____

_____

To further hone and sharpen our reliability leaders, we will enact a practice of annual 360° feedback to build a solid team, working for what Deming called a *constancy of purpose*.

We intend to develop the desired level of leadership and management skills and abilities we expect. We will work with Human Resources and other such agencies to make certain that we have an action strategy to match our strong desire to grow our leadership.

The department leader will find a ready and capable crew eager for solid leadership. An organizational structure will be established to address the department's ability to respond to preventive, corrective, and emergency maintenance simultaneously. The structure will be accompanied with a listing of the necessary staff to make the reliability process work. The staffing members identified will have an accompanying listing of roles, such as:

_____

_____

_____

_____

Planner/Scheduler
Planner
Scheduler
Facilitator
Maintain BOMs
CMMS gatekeeper
PM/PdM overseer
Picklist generator
Data analyzer
Report generator
KPI/Metric generator

Storeroom Supervisor
Mentor
Assistant organizational structure developer
Assistant budgeter
Assistant stock determiner
Assistant report generator
Documenter
Facilitator
Assistant project manager
Supervisor
Stores Stock Committee Chairman

It is the expressed intention of the reliability group to develop a required skills listing and conduct an associated skills gap analysis. This analysis will be complete with an identifying legend and an indication of what the current state is, and what the path going forward will be in order to close the skills gap.

Our skills matrix will look similar see figure on the following page.

Materials Vertical Training Matrix-3/10/16

LEGEND

Level 4
Fully capable of performing on own and training others

Level 3
Fully capable of performing on own

Level 2
Can perform with coaching

Level 1
In training

	Remove and replace gearbox	Rebuild hydraulic cylinder	Align drive belts
John Doe			
Jane Brown			

We plan to seek out and recruit sharp, excited associates in fields outside of maintenance to encourage them to consider joining this team. We believe that our production colleagues have as much interest in reliable equipment as anyone in the organization, and we intend them to have a chance to put their passion to work.

SIX

Working Towards Reliability

"We have talked about the process as a cold thing—
procedures, equipment, flowcharts, and techniques.
The process is brought to life by people.
Our people make the process work; without them, we have nothing."

—Dr. H. J. Harrington, *Business Process Improvement (p. 155)*

Congratulations, you've made it to the last chapter. If you've been using this workbook as a resource, you've no doubt built a very good maintenance and reliability philosophy up to this point. It would be a good idea at this time to actually talk about maintenance. Chapter 6 will include the topics that most of us have come to consider as classic maintenance department responsibilities: preventive and predictive maintenance, planning and scheduling, and storeroom, just to name a few.

I'd like for you to do something just a little different over the next few pages. Try, if you can, to think about these topics as being the responsibility of everyone in the organization, not just the maintenance department. In the case of preventive and predictive maintenance, it is hard to do, as the word *maintenance* is in the title.

Have you ever wondered why it can be so difficult to get production to provide some downtime to perform non-running PMs on the plant assets? It's very simple. It's because our PMs, at least some of them, don't matter. Nothing *bad* happens if we blow past the scheduled inspection. We are going to learn how to give our preventive maintenance some bite in this chapter. We are also going to spend some quality time discussing the entirety of the maintenance work management. But maintenance can't do this alone.

As it was in the beginning of this workbook, so it is here in the closing sections. The care, the reliability, and the maintenance or upkeep of our assets has to be a fundamental responsibility of everyone in the organization. As mentioned earlier, everyone in the organization owes their paychecks to the performance of the company's assets. Shouldn't they have an equally important role to play in the continued survival and performance of those same assets?

Preventive and Predictive Maintenance — Preventing the Inevitable

For some reason most of us get caught up in discussions about what is, or what is not, preventive maintenance. When is a maintenance activity considered to be a predictive activity? If you knew something was going to happen in thirty days, and you took action in twenty-nine days to keep that event from occurring, did you prevent it, or predict it? The answer might shock you. The answer? Who cares? That's the long and the short of it.

Regardless of what we call our process, it's the results that matter. I've heard that the role of the maintenance department is to provide plant capacity, or perhaps the role is to keep the equipment running and to fix stuff when it breaks. Another possible role is one I noted earlier, which is to guard the inherent reliability of the assets. Can all these be true at once? Of course they can and they are.

The way we can do all of this is to have an organizational structure and a guiding philosophy that addresses preventive maintenance, corrective maintenance, and emergency maintenance at the same time. Let's concentrate on that first element for a while.

Preventive maintenance is any maintenance we do to prevent further, more consequential maintenance. And we do this in many remarkable ways.

The Many Types of Preventive Maintenance

If we consider all the differing types of tasks that comprise most preventive maintenance processes, we'd most likely conclude that there are seven distinct *types* of preventive maintenance:

1. Servicing

2. Installation/replacement

3. Calibration

4. Alignment

5. Adjustment

6. Inspection

7. Lubrication

Each type is uniquely different with aspects that make each type significant contributors to any complete PM process.

Servicing
When you think of completing a servicing task on an asset, give some thought to what it takes to get the machine ready for its next production run. Compare that thought to getting your own personal car ready for a road trip. Some activities that you might engage in to get your car ready for a family trip would include:

■ Washing the car inside and out

■ Checking the tire pressure

■ Filling the fuel tank

■ Filling the washer fluid

■ Looking under the hood

Think of some other servicing activities you might perform on your family vehicle prior to a long road trip and record those here:

_____

_____

_____

_____

On a production asset, you might:

- Tighten fasteners

- Clean the machine

- Generally look the machine over

- Restock operating supplies

- Ensure everything is in place (e.g. guards, pallets, etc.)

What are some other servicing activities that might be performed on a production asset at your location? Record those here:

_____

_____

_____

_____

Installation/Replacement

A very common maintenance activity is removing and replacing a component. We've spent some time discussing various aspects of this type of activity, but not solely as a *preventive maintenance* act.

When we replace a component on a hard time (calendar or cycle time) we are actually performing a type of preventive maintenance, though some would consider this a type of predictive maintenance. After all, you could argue, we predicted that the component would fail every six months.

Of course, the component could be a *wear* component such as a belt, a sprocket, or even a wear strip. There are all magnitudes of components that fit into this category of preventive maintenance.

What are some components you install or replace on a regularly identified basis? Record those here:

_____

_____

_____

_____

Calibration

It might be hard to consider calibration as a distinctive type of pre-ventive maintenance, but if you recall, preventive maintenance is a maintenance activity designed to keep more consequential mainte-nance from being required. By maintaining the detailed calibration levels on certain components, we are ensuring that our equipment is running to the established centerlines. This may result in a dual benefit where production is running at its set levels, with a positive effect on product quality, and the maintenance department is keeping the machinery performing in balance.

What are some components at your location that require cali-bration? This could be prior to installing the component, or as a periodic maintenance task to check the unit's calibration. Record your answers here:

_____

_____

_____

_____

Alignment

I'm very much in favor of considering *alignment* as even an alter-nate type of maintenance. This might be referred to as precision maintenance. A reasonable number of folks working their way through this text might recognize this idea. Essentially, if we take care to align components that needed to be aligned, to the degree that is acceptable and within tolerance, then we can be certain that the component will last as long as it is supposed to. As a colleague once told me, it is through precision maintenance that we are minimizing the *chance* of a failure.

Our troubles come when we replace a component through an emergency work order and never get back around to aligning it properly. There's nothing quite as permanent as a temporary main-tenance repair.

There is a competing thought, however, that says that align-ment, or better yet, *checking* the alignment of a component, should never be a PM task. The proper approach is to align the component(s) properly when the item(s) are replaced, and from that point forward, check for vibration or an elevated heat signature.

How is the task of *alignment* used at your location? Please explain here:

_____

_____

_____

_____

Adjustment

This is a classic type of PM, and a classically misused type of PM. Let me explain.

Please highlight any of these that you may have seen or heard before as preventive maintenance tasks:

- Check drive belt, adjust if necessary

- Check roller chain, adjust if necessary

- Check clearances, adjust if necessary

- Check level, adjust if necessary

As you can imagine, these are all horribly suggestive PM tasks. But the intent is in the right direction. We need the applicable components to be adjusted to the proper value, depending on what the component is.

One interesting aspect that almost seems universal in PM tasks that ask for the technician to make some adjustment is that there is never any recording of how much adjustment is left or how much adjustment was made (if it's even possible to gauge that figure). Keep that in the back of your mind.

Inspection

As you might expect, this is perhaps the most familiar type of PM. In this category we are simply asking the technician to *look* at a component and determine if it is in good order or not. As with the adjustment type of PM, the inspection type PM can be very subjective.

Care should be taken to ensure that the person performing the inspection PM is made aware of what is acceptable in terms of condition. Note also, the person performing the PM may be an operator for organizations performing autonomous maintenance.

As an example, list one singular inspection type PM task here. We'll use this for a reference. I've added one as an example:

Check the limit switch for proper operation.

_____

_____

_____

_____

Lubrication

This type of PM is quite literally applying the proper greases and oils as required by the equipment and the lubrication program. Some companies have dedicated oilers. This is an ideal and a highly recommended practice.

As with the inspection type PM, many organizations that have interpreted *Total Productive Maintenance* (TPM) to mean *turning over your lubrication program to the operators* might not be surprised to learn that they end up with a lousy lubrication program.

You've undoubtedly heard lubrication called the *lifeblood* of the equipment. Why would this essential operation be in the hands of anyone less than a seriously trained professional?

Just to recap, there are seven types of PM tasks:

1. Servicing

2. Installation/replacement

3. Calibration

4. Alignment

5. Adjustment

6. Inspection

7. Lubrication

Now that we know that there are several types of PMs, what
do we know of their origin?

Go Straight to the Source

Where do Preventive Maintenance tasks come from? How often
should they be reviewed, or changed, and just as important, who
is responsible for ensuring that our PM/PdM process is on track
and delivering the results we expect? It might not surprise you to
know that we may have lost track of this core element to our
reliability strategy.

As mentioned in earlier chapters, the need for a solid and living
PM/PdM program is fundamental to staving off reactive maintenance. In fact, the antithesis to reactive maintenance is proactive
maintenance, of which preventive and predictive are key.

Listed below are the common sources for preventive and predictive maintenance tasks and requirements. Highlight those which
you and your organization use when developing a comprehensive approach to guarding against failure:

- Original Equipment Manufacturer (OEM)
  recommendations

- Corporate or company policy

- Regulatory (Federal, State, or Municipality)

- Engineering directive

- Tasks applicable to similar equipment or components

- Tribal knowledge

- Good Maintenance Practices (e.g. ASME, NFPA, etc.)

Can you list any other reference sources for PM tasks? If so, list
those in the space provided:

_____

_____

_____

_____

Some of those listed are good and legitimate sources. Taken in context, and with a grain of salt, they form the nucleus of a sound and often successful strategy to prevent equipment failure. We are going to work throughout this section to develop a very relatable and effective preventive strategy and compel others to understand the value of the process, and to be respectful of the fact that we are working for the same end result—more production resulting from more asset availability.

We are going to go all the way back to the beginning regarding the concept of preventive and predictive maintenance and then work forward to build an exceptional methodology. As such, consider the role of maintenance and how the proper execution of maintenance activities can lead to greater productivity.

Preventive Maintenance Contributes to Greater Productivity

What is the role of maintenance in your facility?

_____

_____

_____

_____

Most would write, as we did before, 'to fix stuff' or 'to keep the plant running.' More formally, the role of maintenance should be to guard the inherent reliability of the asset. Equipment arrives at our facility with an inherent reliability. As has been pointed out in many books on the subject, you can't make an asset more reliable than it was intended to be originally. No amount of corrective or preventive maintenance is going to make an asset more reliable. If it's designed poorly, built cheaply, and installed hastily, you won't have a world-class asset. You will more likely have a nightmare on your hands for the next forty years.

The role of maintenance in your institution should be, primarily, to guard the inherent reliability of the assets. In that service, you are pressed to determine the best course of preventive, predictive, corrective, and yes, reactive maintenance to keep that asset operating to the benefit of the organization.

We need to stop right here to discuss something you may never have thought about. Imagine for a moment that you are responsible for production at your location. In this example, I'll be the person who is responsible for asset reliability and maintenance. One day I come to you and try to convince you that, if you'll stop operating your machine in order for me to perform my PM checks, I can assure you of greater productivity. In short, if you stop regularly, you'll make more product. How do you convince someone that they can make more by running less?

That is a tough argument. By the way, production will run more consistently, and that's the secret to more productivity.

A Practical Example

Here is a basic example. Let's say that you can drive your family sedan about 400 miles without refueling. Said another way, your car can travel 400 miles before the engine dies due to lack of fuel (this is a failure mode). You can prevent that by stopping to get gas before your car runs out of gas. With that simple strategy, you might be able to drive indefinitely. But we know that's not completely true because the owner's manual and our tribal knowledge about vehicle care and operation tells us that there is more to having a car operational than just putting gas in the tank.

Through additional preventive maintenance, condition based monitoring, and condition based maintenance (often referred to as predictive maintenance) we can drive, seemingly, forever.

Table 6-1 might reflect the comprehensive preventive and predictive approach on your family car:

Table 6-1: PM strategy for the family sedan

Milestone	Activity
Every 350 miles	Get gas
Every 5000 miles	Change oil, rotate and check tires, refill fluids, change air filter
Every 50,000 miles	Replace tires based on condition
Every 100,000 miles	Overhaul

The result is a plan that requires maintenance to be performed by a professional and some work that is considered autonomous. The need for all this work is arrived at by reviewing all the sources previously listed at the beginning of this section. By working to prevent the failure of the family sedan, you are actually working to prevent the consequence of the failure. The consequence of failure of the family car is inconvenience at best, and could be fatal at worst.

With the maintenance plan we just created for the family vehicle, we are stopping the car based on a predictable schedule, and the car ends up running more consistently. Does that sound familiar?

Coincidently, plant equipment only fails while in use. That sounds silly, but haven't you heard someone say, "It broke in the middle of production?" There is a consequence to failure, and the prevention of this failure is our ultimate aim. The more we understand about failures, the better able we are at avoiding them altogether, or at least holding them off for as long as possible.

How do we do this? The P-F curve.

The P-F Curve

There are many wonderful works in circulation regarding the P-F curve and this brief introduction is not meant to rewrite what you know to be true, but simply to use as a base for understanding where preventive and predictive maintenance fit. This is likely a review for most, but for some it will be new and valuable information.

Most components—not all components, but most—follow a predictable deterioration curve, and that is the P-F curve, as shown in Figure 6-1. If the component in use is the right component, installed correctly, aligned, and cared for properly, that component will last a reasonably long time. The unfortunate nature of the industrial world is that most components don't make it to the end of their natural lives.

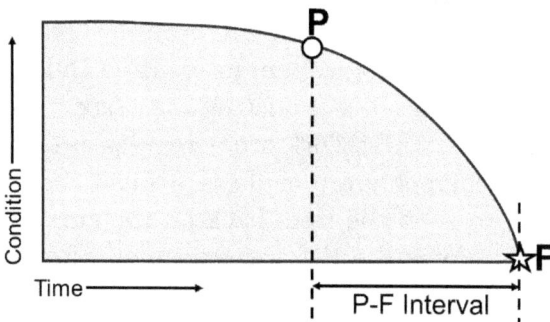

Figure 6-1: P-F curve

Figure 6-1 is used to demonstrate the typical deterioration curve of a sample component. The point *P* marks the point of potential failure. This is the point on the curve that people, using only their own senses, can notice a change of state.

For example, an operator might notice that a roller bearing was running hot, but it was not running hot last week.

On our P-F curve, the point *F* marks the point of functional failure. The component doesn't actually have to stop working at the point *F*, simply one of its functions has to have failed.

The P-F curve gets its name from the interval of time that is present between the point *P* and the point *F*.

Figure 6-2 shows the relationship that preventive and predictive maintenance have with the P-F curve.

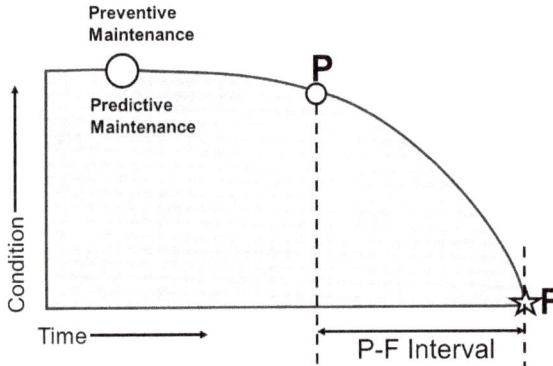

Figure 6-2: PM/PdM as it relates to the P-F curve

To be clear, in Figure 6-2, the location of the PM/PdM nodule isn't meant to represent that this is the only location on the P-F curve that we conduct proactive maintenance; rather, this point is meant to indicate where we hope to *keep* the component in terms of deterioration through excellent preventive and predictive maintenance. I call this *keeping it back on the curve.*

If, for example, a PM work order results in the immediate replacement of a component, it could be argued, and rightly so, that the PM failed to keep the item back on the curve. This is not an effective PM because the PM was executed after the component had progressed up to the point *F* on the curve.

For preventive and predictive maintenance to be effective, it has to result in the component 'staying back on the curve.'

When discussing preventive and predictive maintenance it is important to note that we want proactive maintenance that is both effective and efficient. In terms of PMs/PdMs, being effective means having a PM that 'sticks.' As an example, when a technician completes a *thirty-day* PM, their signature indicates a professional opinion that the component inspected or involved will last at least another *thirty-one* days. This is an effective PM.

What makes a preventive or predictive task efficient? This is a combination of who performs the task (maintenance or operator), how quickly and completely it is accomplished, and whether the PM tasks (steps) are constantly reviewed to keep making the inspection more thorough and complete.

What follows would be a typical observation in almost any factory or facility. Figure 6-3 is a typical factory layout. The rectangle shows the walls of the plant; the squares are the different assets in the plant.

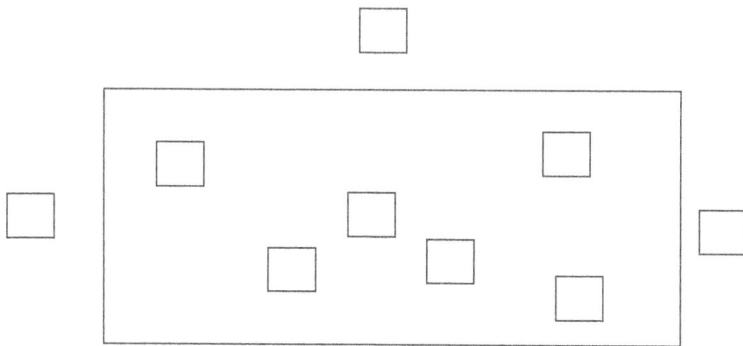

Figure 6-3: Typical plant layout

Figure 6-4 illustrates that same plant layout with black dots representing each and every PM or PdM task performed on individual components throughout the year. In a year's time, these black dots come to represent every activity that was conducted on the pretense of keeping failures at bay, and more importantly, every consequence of those failures in check. The collective wisdom of all our research and the avalanche of sources clearly spells out, "This is our proactive strategy."

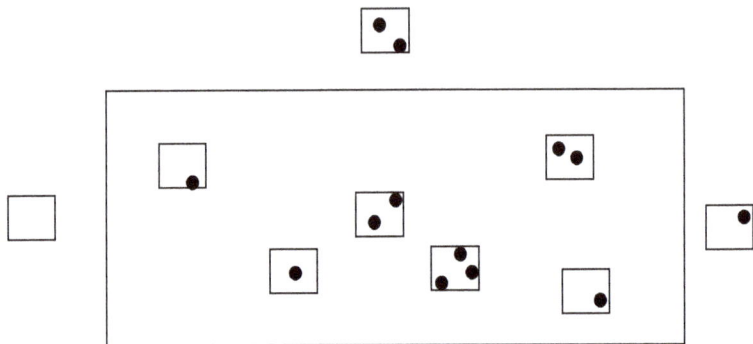

Figure 6-4: Every component "touched" by a PM or PdM action

It we superimpose onto this more recent layout a circle that represents those component failures experienced in a two year time span, we're left to conclude that it is the very components we are touching through inspection that are failing. This is shown in Figure 6-5.

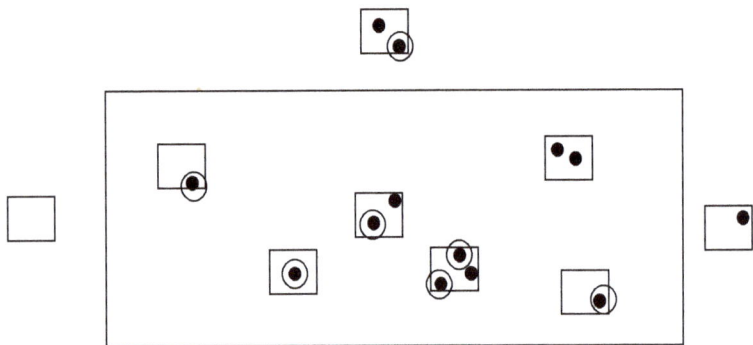

Figure 6-5: Plant layout indicating component failures

It is almost with precise accuracy that we can guarantee that there has never been a component failure in a plant or facility that doesn't have a PM, PdM, or at least an operator check associated with it. In fact, the certainty goes beyond that with a prediction that the very next utterance from the plant manager is, "Who PM'd that last?"

This is a classic representation of the ineffectiveness of our PM program. Let's face it, our PMs just don't *stick*. In order for a PM to *stick*, the component has to last through the next scheduled inspection. That means when a technician signs their name to a PM work order, signing off a monthly PM on a motor, that motor is guaranteed to run for the next thirty-one days, barring any catastrophe. Now that is an effective PM!

Complete Table 6-2 by listing the last five component failures you've experienced in your plant or facility and listing the PM/PdM tasks that were meant to address the very component in question.

Table 6-2: Component failure vs. PM/PdM task

Component	Task

For those components you have listed, and for the components circled in Figure 6-5, it would be accurate to say that the PMs for these components did not *stick*. But where did it all go so wrong? It could be that it was never right to begin with.

The Source is the Problem

Earlier in this chapter we covered some sources that would be good resources to use when developing the PM strategy for an asset. It is becoming less and less likely that the OEM can serve as a reasonable source for PM mandates. In fact, some OEMs are just *phoning it in* to borrow a phrase, meaning that they (the OEM) are not even trying.

What follows is the actual suggested PM protocol for an asset a client purchased. This machine cost $250,000, and the client purchased twenty-seven of them at one time for a total of $6,750,000. Again, this is what the client received for the investment in terms of a suggested maintenance program:

In addition to performing a daily, prior to use inspection, there is also a series of monthly and biannual maintenance procedures to be performed on the system.

Every month

- Check for pinched hoses

- Check for air leaks

- Replace worn hoses

- Tighten loose fittings

Every six months

- Check for loose or missing fasteners and repair or replace

- Grease all fittings

- Grease all pins

- Grease all trunnions

That is it!

Consider again this minimalistic approach an OEM took to a multi-million dollar, turn-key project. The OEM was paid to give sage and professional upkeep advice, yet delivered neither. We should be in agreement that the PM strategy laid out by any OEM should just be a starting point to keep all of an asset's components *back on the curve.*

In the end, the OEM did have some very salient views on what proper and good maintenance should be. The same document just referenced continues to explain that preventive maintenance must be a process of systematic care to include inspection and servicing of equipment. This work is done in order to maintain the equipment in serviceable condition as well as detect and correct early and minor failures before more costly and time consuming repairs or replacements are required.

This particular equipment manufacturer builds a strong argument for the proper operation and operator care of the equipment. Operation and operator care contribute to preventive maintenance as much as the maintenance inspections and services. It is often adverse conditions such as extreme temperatures, moisture, or dust that cause equipment to require more servicing.

The OEM, being referenced, adds that an effective planned maintenance program has four categories:

1. Planned periodic inspection
2. Adequate and correct lubrication programs
3. Corrective adjustments and repairs
4. Complete and accurate records for historical and trending purposes

Planned Periodic Inspection

Inspections by qualified and trained personnel are designed to detect trouble before it happens and to schedule required repairs quickly, without unnecessary disruption of production. Inspectors should be thoroughly familiar with the operation of the equipment. The inspection method may vary from a walking visual inspection to an actual dismantling check at certain points to determine wear or other failure attributes.

Adequate and Correct Lubrication

It is impossible to overemphasize the importance of the correct lubrication of any piece of machinery or system component. Lubrication is singularly the most important work that maintenance accomplishes in terms of preventive efforts. Unfortunately, the subject of lubrication does not receive the attention it should in many plants or supplier training programs.

Dr. Nathan Wright gave a boost to the importance of lubrication in a reliability program in his work, *The Death of Reliability*. "Reliability is an organizational level effort and to achieve this the organization must address the three main root causes that result in unreliability: improper lubrication, contamination, and improper installation. Of these three, lubrication, or better said, improper lubrication is the most prominent." (p.14)

Correct Adjustments And Repairs

Inspectors should work from a planned checklist, explained the equipment manufacturer that we are using as our example. This checklist should cover all points of wear on the components, and have a column indicating each task is 'OK' or that the component required additional maintenance, such as repair or adjustment. That column might look like Figure 6-6:

OK	ADJ	CM

Figure 6-6: Blocks to mark at the end of a PM task

This is the point at which the training, experience and good judgment of the inspector is of prime importance. Outside of detecting the need for immediate action from these reports, changes can be made in the lubricating schedule, and adjustments and repairs or replacements can be scheduled ahead for planned downtimes.

Complete and Accurate Records
The importance of maintaining complete and accurate records should be evident to the audience reading this workbook. It is from these records, this particular equipment manufacturer educates us, that the *"experience* factor can best be determined; repairs and replacements scheduled at convenient times in advance of the development of problems."

These records also will aid in determining the type of spare parts that should be kept on hand or ordered in advance of a major repair. These records can also be used to avoid the overstocking of parts which are not required, thus helping reduce dead or slow moving inventory.

A preventive maintenance program patterned along the foregoing lines, but tailored to the needs of a specific plant, will keep your plant running more cost effectively in the long run.

The OEM that we've been citing has done an admirable job in explaining the benefits of a comprehensive preventive maintenance program. However, the juxtaposition of the actual suggested proactive program they offered as shown at the beginning of this section is puzzling.

How can an OEM develop a spot-on preventive maintenance protocol for an asset in any plant, if the OEM didn't build a single component on the machine?

OEMs Don't Manufacture Much, Not Really

The company that built your equipment, even if it was your own company, didn't build any of the components on the equipment. They bought the components just like everyone else, and attached them to the frame of the machine that you were buying. The manufacturer of those components has no idea of the environment in which you intend to subject them to.

The following information is not meant to be patronizing, but informative. An OEM is only going to provide enough preventive maintenance information to comply with your scope of work that requires them to 'provide a recommended preventive maintenance schedule.' The money for the development of the PM schedule was not in the OEM's bid or, quite frankly, in their budget. They worked to prepare and present the most cost-effective and successful bid they could. They did not budget hours upon hours of engineering time to develop a PM program they knew quite well you wouldn't follow anyway.

That last comment was kind of harsh but considers the fact, as previously mentioned, that the OEM doesn't know your applications as well as you know them. Without this knowledge, they are unable to prepare a comprehensive preventive maintenance plan for your equipment. That much should be evident. The best the OEM can do is to make recommendations for general checks and provide the cut sheets for the components on your machine. It is up to you to take the cut sheets, knowledge of your application, and the overall general recommendations and develop the PM strategy.

It may seem odd to say, but in the spirit of turning lemons into lemonade, we may actually be well served by a less than stellar PM strategy from the OEM. We can build our own if we know how.

PMO

Our approach towards preventive and predictive maintenance needs to align with the history of our equipment's performance. This is most recognizably accomplished through *Preventive Maintenance Optimization* (PMO). This requirement is valid for preventive maintenance steps for brand new equipment, as well as for PMs on equipment that has been operating in our plants for years.

Because we know that PMs are generally created by a less-than knowledgeable OEM, or by our efforts to keep from having repeat failures, we can surmise that our PMs do not generally reflect our best effort. In fact, it is reasonably understood that we think we're doing our best work. We most likely have not applied any level of *critical thinking* to the development of our preventive maintenance strategy.

That's a tough truth, and even harder to accept since we now know that a good PM strategy is one of the elements needed to get us out of a reactive mode. In a roundabout way, we've actually done more harm than good in our efforts to prevent equipment failures and may have added to the consequence of failure.

PMO is sometimes mistakenly seen to be in competition with two other, more well-known reliability methodologies: FMEA and RCM.

If you were going to build a new plant on a cleared piece of land, you'd be correct in using the methodology of FMEA (*Failure Mode and Effects Analysis*) to develop a maintenance and reliability strategy. This would be a *green field* exercise.

If, instead, you were building a new asset, a new piece of equipment, you'd correctly employ RCM (*Reliability Centered Maintenance*) as your approach. This would be a *white paper* exercise.

If, on the other hand, you have an existing plant with existing equipment, the method you might want to consider to improve your existing reliability and maintenance efforts would be PMO.

As with FMEA and RCM, PMO essentially seeks to find the failure mode and determine the best approach to take to keep the failure mode from occurring.

In RCM and FMEA we typically look at all the various ways equipment can fail and work to find ways to identify what causes the failure and minimize its effects or eliminate the potential all together. RCM and FMEA are great practices, of that there is little doubt or debate, but they require tremendous resources to do correctly and are often overwhelming events.

PMO, though, is a shorter process, netting virtually the same results as RCM. However, one significant factor beneficial to PMO is that we don't concern ourselves so much with 'what could fail;' we focus instead on 'what has failed.' Since we are concerned with equipment that has been in operation in our plant for years, we are concentrating on the failures that equipment has suffered over time and working toward avoiding repeats of those exact failures.

The PMO process is conducted as a *workshop* meant to create an energy that only a team can generate. Although PMO is a systematic process, this team can really expand the boundaries of conventional thinking and standardized workflow.

As Liker recorded in *The Toyota Way: 14 Management Principles From the World's Greatest Manufacturer*, "... standardized work was never intended by Toyota to be a management tool to be imposed coercively on the work force. On the contrary, rather than enforcing rigid standards that can make jobs routine and degrading, standardized work is the basis for empowering workers and innovation in the work place." (p. 142)

In the PMO process, we are going to use all data at our disposal. Hopefully much of that data will be fact-based from our CMMS and/or other records such as operator log books, run reports, etc. We also want to include interview material with maintenance personnel and operations personnel. Essentially, we want to include all manner of information in our 'optimization' of the PM. We even want to review the parts-checkout history from the storeroom relative to the equipment we are reviewing.

In Table 6-3, list locations where equipment history might be found in your facility. We've mentioned some already, such as the CMMS, and storeroom records. Where else is information, formally or informally, stored at your site?

Table 6-3: Locations for equipment history

Once we collect and separate the data into useful information, the review can begin. We can best see what we hope to achieve in the simple Figure 6-7:

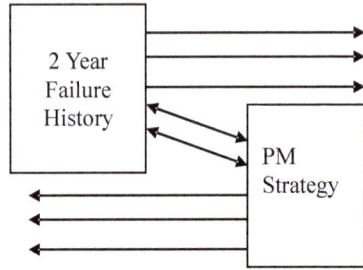

Figure 6-7:
The general approach to PMO

Figure 6-7 demonstrates three conditions (reference the arrows in Figure 6-7):

1. There are many failures in our history for which we have no preventive or predictive strategy.
2. There are some failures for which we have seen defects and failures, yet we have a preventive strategy.
3. There are many PM tasks performed for which we've never seen a defect.

For those instances where we have experienced failures and do not even have a preventive strategy against the failure, we've generally overcompensated by adding several PM tasks to our program in an effort to cover our rears (CYA). We don't ever want the situation to repeat. We overcompensate by violating one of the golden, but oft forgotten rules of reliability maintenance: one failure mode = one preventive task. No more, no less. This contributes to the thesis that 20%-40% of our PMs are a waste of time. We attempt to pile on several PM checks against a common failure mode, and the component still fails (see Figure 6-5).

Example: A petroleum refinery in Europe processes gasoline for consumption in North America. In an effort to transfer 'ownership' of the equipment reliability, all the preventive maintenance tasks were moved to operator-performed tasks. This was done prior to any PMO effort. The company moved (and this is a real number) 920,000 annual PM tasks to the operator corps. During a PMO review of a single pump in the refinery, here is what was discovered:

The pump operates a total of two hours per week. It is a small transfer pump and continuous use is not required. The operators perform a daily predictive task on the pump using a small vibration pen to monitor and record the vibration level on the pump. The vibration reading is taken every single day, regardless of the pump's operation; running or not. Once a week, a professional contract service is on site to take a vibration reading as part of a plant predictive maintenance program. On many occasions, it has actually happened that the operator and the professional vibration analysis monitor were taking readings at the same time, on a pump that was not even running.

I was there; I saw this happening.

PMO helps us to align what is failing to what we are doing about it. We are going to enhance our PM/PdM strategy to cover the failures we've experienced as a means to guard against them recurring. Likewise, we are going to eliminate the PM/PdM tasks for which there never has been a defect and/or, if there ever were, it would be inconsequential.

For those items that fail, and we have preventive measures against such failures, we are going to nail down the failure mode and arrive at the best approach to prevent those failure modes.

Our preventive strategy will be *optimized* to look like Figure 6-8.

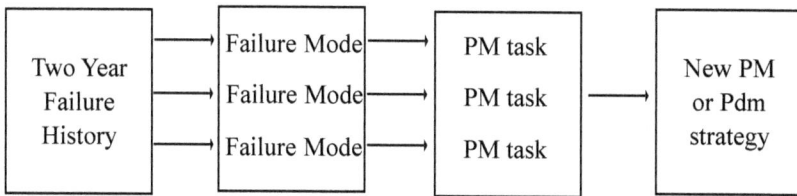

Figure 6-8: New PM or PdM strategy

Understanding The Failure Mode Makes The PM Task Clearer

We are going to close out the section on preventive maintenance by exploring some failure modes and, through doing this, write a much better PM task. Our new task will be better defined, better explained, and much clearer.

Keep these five principles about a PM task in mind. A well written PM should instruct the reader on:

1. What to do

2. How to do it

3. What good looks like or, what the acceptable criteria is

4. What to do if it is bad or unacceptable

5. Any safety considerations to take before performing the task

From the following examples, see if you can identify the five principles of a PM task in each.

Example 1

Old PM task:	Lubricate top bearing 4 grams
Failure mode:	Bearing seizes due to insufficient lubrication
New PM task:	While machine is running, with a clean rag, wipe grease zerk clean; apply 4 grams of Mobile XHP222 grease using calibrated grease gun; with a clean rag, wipe grease zerk clean (bearing part #ZO00002407)

Notice that I put the part number of the relevant part in the PM task. I do this so the planner can more easily create a corrective maintenance work order if necessary.

Example 2

Old PM task:	There was none, but a failure history on the grease lines did exist
Failure mode:	Bearing seizes due to insufficient lubrication
New PM task:	Visually inspect grease line to top bearing looking for signs of crimping, cracking, or other evidence of line compromise. There should be none. If line is damaged, notify maintenance lead. If line is not attached properly to top casting and the grease zerk at base assembly, reattach.

Example 3

Old PM Task:	Clean motor coupling
Failure mode:	Motor coupling fails due to buildup of product between flexible coupling and driven flange

New PM task: Lock out and tag out the equipment before approaching the assembly. Using a scraper device, or a can of compressed air, clean the gap between the flexible coupling and the driven flange. There should be a 1/8 inch gap between these two surfaces. Use a feeler gauge to confirm the gap in four different places around the coupling. If the gap cannot be cleaned, or the gap is in excess or less than required, notify the maintenance lead.

For the last example, answer the following questions:

What was the reader being asked to do?

_____

_____

How was the reader to do what was asked?

_____

_____

What was the acceptable criteria?

_____

_____

What was the reader to do if the task could not be accomplished favorably?

_____

_____

What were the safety precautions?

_____

_____

Condition Based Maintenance (CBM)

Although we've spoken almost exclusively about preventive maintenance during this section, it is important to note that predictive and condition based maintenance are additional elements of a solid proactive maintenance program.

I've had many discussions over the years centering on the place for and the meaning behind CBM or *Condition Based Maintenance.*

I would like add some thoughts to this and I feel we owe it to ourselves and our organizations to build a process that works for our locations, despite some of the differences in the terms we use.

On its face, CBM clearly becomes the vehicle by which we introduce technology into our abilities to *sense* conditions back on the P-F curve. As Suzuki explains in *TPM in Process Industries*, "Condition-based maintenance uses equipment diagnostics to monitor and diagnose moving machinery conditions continuously or intermittently during operation and on-stream inspections (OSI—checking the condition of static equipment and monitoring signs of change by nondestructive inspection techniques)." (p. 148)

However, we're making a rookie mistake to compare CBM to PdM.

I would suggest that we move beyond the common belief that, "The terms *Condition Based Maintenance (CBM)*, and *Predictive Maintenance (PdM)* are used interchangeably." (Gulati, *Maintenance and Reliability Best Practices*, p. 52)

The trouble with this short-sighted comparison is that it leaves an incomplete strategy in play.

Here is an example to explain this point.

Joe Jones, a Maintenance Manager at a northwestern manufacturing company compelled his plant leadership to allow him to set up a contract service to perform some predictive maintenance. Joe had attended a reliability seminar and heard several testimonies on how the application of technology had helped save thousands of dollars and advance other reliability programs in various industries.

Like most maintenance leaders, Joe located a local contractor for the work. The contractor performed a plant tour and suggested a series of vibration, infrared, and oil analysis testing. The results of each test would be communicated on the contractor's website, under an exclusive log-in for Joe and his team. The contract service would be executed quarterly. This was to be an $80,000 a year contract.

The first few quarters of the contract service ran smoothly. Everyone, mostly Joe, was pleased with the thoroughness of the program, and the production equipment had never run so well. This was great.

During the third quarter assessments, the contractor discovers a motor that is reading at a vibration level beyond what is considered 'normal' and proper. The contractor feels strongly enough about the readings that he actually reports immediately to Joe. Joe is concerned and the general recommendation is that the motor should be replaced.

Here is the problem.

Immediately half of the maintenance department is in lockstep agreement with the order to replace the motor, because, after all, the numbers don't lie. The other half? Well, they thought predictive maintenance was all smoke and mirrors to begin with. Just another way for a contractor to rip off the company.

This may describe how predictive maintenance works at your location.

Condition Based Maintenance is not the same as Predictive Maintenance. Condition Based Maintenance evokes the idea of Condition Based Monitoring.

Condition Based Monitoring is the activity of monitoring the *condition* of the asset, often bringing in the technologies that predictive maintenance brings to a proactive process. I would argue that it's *Condition Based Monitoring* that is synonymous to Predictive Maintenance, and not Condition Based Maintenance, as suggested earlier.

In his edited work, *TPM in Process Industries*, Suzuki continues in his quote from earlier by stating that, "As its name implies, condition-based maintenance is triggered by actual equipment conditions rather than the elapsing of a predetermined interval of time." (p. 148)

Where does that leave us?

Predictive Maintenance (PdM) is a part of a Condition Based Maintenance program. In either of the two processes, once a certain condition is noted, the condition *triggers* a response. That *response* is the Condition Based Maintenance.

Here is the example of that in action. Remember the scenario from earlier. Joe Jones has set up a contract predictive maintenance service with a local outfit. That crew performs quarterly assessments of the equipment that was set up during the initial round of assets to monitor.

In the third quarter, the PdM contractor discovered a motor with a troubling vibration reading. The contractor suggested that the motor should be replaced. We left the scenario earlier with half the maintenance team in agreement, and the other half opposed to this suggestion.

True Condition Based Maintenance should include *predetermined steps* to take, on the basis of the condition of the component.

Joe and his entire team would have agreed to the following maintenance plan to execute on the condition of a high vibration reading on this particular motor. Those procedures may include these steps, in this order:

1. Record the bearing temperatures at the front and rear bearing locations.

2. Confirm that motor is securely mounted; if vibration is not reduced, then...

3. Apply a shot of grease to each motor bearing; if vibration is not reduced, then...

4. Shim motor as may be needed; if vibration is not reduced, then...

5. Check motor alignment and adjust as needed; if vibration is not reduced, then...

6. Replace the motor if the reading is not reduced.

Imagine that Joe's PdM contractor had been taking infrared readings and had come across a roller bearing reading at a very high temperature. Using the example of establishing a condition based maintenance plan that would be triggered by the condition, what series of steps would you suggest to take if you had Joe's confidence? Record your thoughts here:

_____

_____

_____

_____

We've spent this time developing some thoughts on proactive maintenance primarily because this is where we need to be—back on the curve. In fact, in a very significant way, we could argue that the P-F curve starts at the PM/PdM nodule. That isn't scientifically true, but for practical purposes, it is as true as it gets.

We are going to continue throughout this chapter to build on this P-F, so it makes sense that we start at the beginning, back on the curve.

You Have To See It Before You Can *See* It

Recall that the point P on the P-F marks the point of potential failure. This is the point on the curve that people, using only their own senses, can notice a change of state. This nodule may also be referred to as the change of state point.

For example, an operator might notice that a roller bearing was running hot, but it was not running hot last week.

Once a component has been detected by someone, using their senses, and deemed to be of concern, the entire maintenance operation kicks into high gear. This is the point at which we can *see* the failure manifesting itself, and we can take action. At this specific point, we have *identified* an issue. Figure 6-9 shows an example of when an issue is identified on the P-F curve.

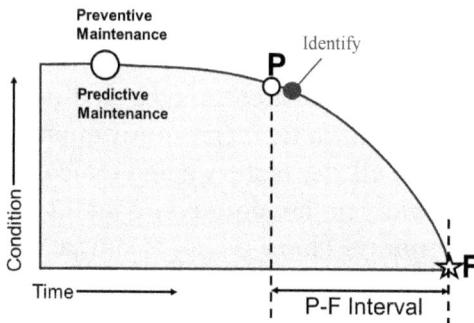

Figure 6-9: Identify nodule on the P-F curve

I left the PM/PdM nodule on the P-F curve in Figure 6-9 simply to communicate the importance of maintenance identifying a failure or failure mode way back on the curve. By identifying the issue back on the curve, the entire organization has more time to develop a plan and a planned corrective action.

In a company that practices Total Productive Maintenance, I would encourage maintenance to work in the back area of the curve, and I'd encourage the production folks, with their autonomous maintenance, to get really good at identifying an issue as close to P as possible.

There are some commonly accepted rules around identifying issues that bear mentioning. We want everyone in the organization to have the ability to identify and report anything that is not correct or that they have a concern with. Consider the associates as your *eyes and ears* as to what is really happening at your location. Anyone in the organization should have the ability to request work from the maintenance department.

A few years back I was working on a facilities maintenance assignment at the enormous campus of a major company. The campus was downtown, and consisted of one twenty-seven story tall building, and two fourteen story tall buildings. Tying these complexes together was an equally impressive mezzanine facility. Again, this was a very large enterprise.

When I got to the campus, and found the facilities engineer (my host for the assignment), I asked him, "What am I doing here? What do you hope to get from our work together this week?"

He said that he wanted his guys to be more *proactive*. When anyone (clients or a consultants) says that they want to be more proactive, you need to challenge them on what that word means.

I asked him what he meant by proactive. He continued to explain, telling me that in facilities maintenance, proactive means that his guys catch all the facility's defects before the users of the facility do. This was one headquarter's facility and everyone in it was with the company. This was not a contract maintenance crew.

He said that he wanted his guys to catch any stained ceiling tiles before someone in accounting found them. His maintenance team should find a leaking toilet valve before an associate in human resources noticed it. A maintenance team member should discover a sticking door latch before the folks in transportation came across it.

I resisted the urge to tell him that his definition of proactive isn't the correct definition. Instead, I told him that his was an admirable goal. I asked him how many maintenance people he had.

He said seven. I asked him how many people worked at the location; the answer was twenty-seven thousand. Yes, you read that number correctly.

I wasn't a math major in school, but I did take some math. I did a quick calculation in my head and told the facility engineer that seven divided by twenty-seven thousand equals "you lose."

I told him that he really doesn't want his guys to find stuff before the facility users. Rather, he should deputize all twenty-seven thousand associates to 'be on the lookout' for issues and report them promptly and properly.

That is what we really want. Believe me, we not only should want our people to be on the lookout for issues, but in today's competitive businesses, we absolutely *need* everyone to be engaged in what's going on.

In the space provided, list the process steps required for an hourly associate to identify and report an issue through proper channels at your location:

_____

_____

_____

_____

In order for a machine operator, or even a maintenance technician to properly identify an issue that is a concern, or to use our example, an issue that is a change of state or point of potential failure, they (operator and technician) need to know the theory of operation of the equipment.

If you don't know how the machine is supposed to work, how would you know if it isn't working correctly?

When a concern is forwarded, it needs to be through the proper channels and methods. It might be wise to ask what *proper* means. Remember the chapter discussion on process guides. The proper channels and methods are *whatever your team says they are.* Simply have a formal system that works for your organization.

Just because an issue has been identified and properly reported, that doesn't guarantee that anything will be done about it, or done any time soon. The next step to be taken for an issue to be recognized and attended to is to set the proper priority.

What Is The *Real* Priority?

The requestor sets the initial priority; this is an absolute must. When a requestor is asking for assistance from the maintenance department, that person (the requestor) must communicate how important that need is. It should be understood that at this point we are only talking about how important the requestor feels the need is. In the larger scheme of things, the requestor's need may be very low in priority.

If we allow work requests, or work order requests to be submitted without an initial priority from the requestor, then is maintenance left to simply *guess* how important the need is?

Figure 6-10 is meant to indicate the approximate location of this initial priority setting on the P-F curve. Of course in practice, the initial priority occurs when the work request is made. In that case, the priority setting is instantaneous with the work request.

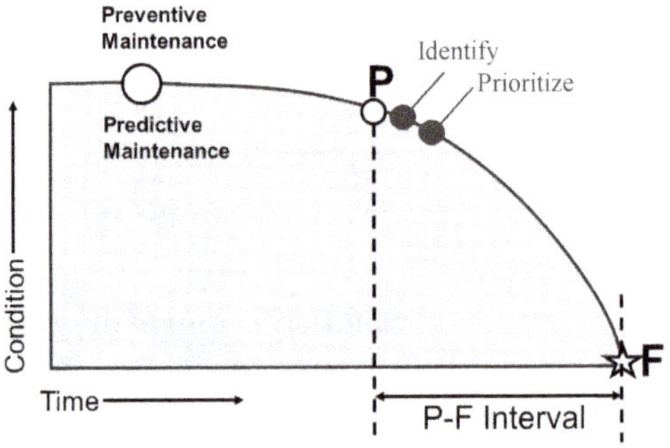

Figure 6-10: Prioritize nodule on the P-F curve

In the next section regarding the evaluation of a work request in the approval process, we will discuss who gives the work its real priority. One thing is for certain; in order for the requestor and the approver to have a chance of arriving at the same conclusion as it relates to priority, these two folks need to have the same reference point.

Does it RIME?

A common method used to establish priority in a plant or facility operation is to compare the importance, or criticality of the asset in question, with what the maintenance need is at the time. One of the most common tools to do this is the RIME chart. RIME is an acronym for *Ranking Index of Maintenance Expenditures*. This is the chart that lists assets down the left side, in order of importance, and the maintenance codes across the top, in the order of severity.

Figure 6-11 is an actual RIME chart to see as an example of what one might look like. Weights for each category of asset criticality and maintenance activity code are given, starting with 10 as the highest. The numbers in each cell are the multiple of the row and the column weight.

Rank	Equipment	Breakdown Critical Safety	PMs	Corrective	Service	Routine	Facilities and Grounds
Rank		10	7	6	5	2	1
10	Main Power / Air Compressor / IT Network / Furnace	100	70	60	50	20	10
9	Main Water / Cutting Table 6 / Line 1 / Line 2	90	63	54	45	18	9
8	Cutting Table 8 / Line 8 / Line 00	80	56	48	40	16	8
7	Line 3 / Line 7 / TTUM 2000 / STM	70	49	42	35	14	7
6	BND 17 / Cutting Table 4/5 / MUTT 4000 / Hawkeye	60	42	36	30	12	6
5	Batch Saw / Filler / TFB Bender / Sling Crane	50	35	30	25	10	5
4	Pattern Area / Chopper / Spider	40	28	24	20	8	4
3	Forklift / Sweeper / Docks	30	21	18	15	10	3
2	Trash compctr / Rec compctr	20	14	12	10	4	2
1	HVAC	10	7	6	5	2	1

Priority 1

Priority 2

Priority 3

Figure 6-11: Sample RIME chart

When putting a RIME chart together, the first course of action is always to agree on the criticality of the assets. Everyone, including maintenance, operations, engineering and others need to agree to the order of the equipment's criticality.

At your location, what asset is the most important asset?

Without any coaching, ask three other people in your organization what they believe is the most important asset. To give this survey some credence, ask three folks outside of your department. Record their thoughts here:

_____

_____

_____

_____

For the RIME chart in Figure 6-11, we still need to add some definitions.

The RIME chart we are using as an example uses the following definitions for maintenance activities:

- Breakdowns/Critical Safety—A major production asset is down, or imminent danger exists

- PMs—A non-running preventive maintenance task is required for continued operation

- Corrective Maintenance—Planned work to correct noted deficiencies

- Service—Other related tasks to keep assets in service

- Routine—Unplanned work, not critical

- Facilities and Grounds—Cosmetic and/or non-emergency work in support areas

With a little effort and some vocally expressed feelings, the group previously assembled will have no trouble arriving at the arrangement of assets into a criticality list. This is primarily because that team will derive some sort of approach to use to begin with.

It would be very typical to have breakdowns and safety as the most *weighty* of the maintenance activities. I'd even go so far as to say the order is more likely to be safety and then breakdowns. The rub lies in the definition of those usually well understood terms.

Here is an example: Using the RIME chart provided as an example, what maintenance category do the following concerns fit into?

A missing hand rail? _____

A loose hand rail? _____

A rusty hand rail? _____

A bent hand rail? _____

Priorities 1, 2, and 3 also have to have some interpretation. For the example provided in Figure 6-11, the following definitions are true:

- Priority 1—Failure to perform work *will result* in death, severe injury, significant damage to company equipment, or major loss of production

- Priority 2—Failure to perform work *may result* in injury, damage to company equipment, or loss of production

- Priority 3—Failure to perform work may result in loss of equipment integrity, adherence to reliability procedures, or general order and discipline

Using the same hand rail example above, and given the priority definitions provided, what priority would you give each scenario?

A missing hand rail? _____

A loose hand rail? _____

A rusty hand rail? _____

A bent hand rail? _____

The description of a priority should include:

1. The definition of that priority, in exact terms

2. An example of that priority actually occurring in the facility

3. The response maintenance is to take

Using the RIME chart in Figure 6-11, and our discussions over the last few paragraphs, Figure 6-12 is an example of what a priority 1 might look like:

```
┌─────────────────────────────────────────────────────────┐
│                        Priority 1                         │
│  Definition—Failure to perform work will result in death, severe │
│  injury, and significant damage to company equipment, or major loss │
│  of production                                            │
│                                                           │
│  Example—Line 7 infeed cylinder fails to retract          │
│                                                           │
│  Response from maintenance—Work will commence immediately │
└─────────────────────────────────────────────────────────┘
```

Figure 6-12: What a priority 1 means

Why do you think the phrase *commence immediately* is used in the maintenance response in our sample definition for a priority 1? Can we always correct an emergency condition? No. But we can always run a maintenance technician out to the machine to see if it can be brought back up and running. This could buy us time to develop a permanent repair and plan the corrective work.

At your location, what does a priority 1 indicate? In the space provided, record the definition, a real world example at your location, and the response that is required from maintenance:

Definition: _____

An example: _____

Maintenance's response: _____

After the Priority is Agreed to, Everything Else Flows Smoothly

It is important to mention that this line of discussion is meant for work that can be planned and scheduled. If you give the P-F curve a glance, you'll agree that for emergency, or break-in work, the component would already be a point on the curve approaching or past the functional failure, or *F*.

Read this carefully, because this is truly important. Your work request process will succeed or fail based on what we are about to discuss.

It is vital—no, make that absolutely *critical*—that all work entering your maintenance and reliability processes come in on one path. Again, this discussion is only concerning non-emergency work. The work request process is the only route by which work can enter the maintenance process from outside of the maintenance organization. This goes for anything the plant manager may need as well.

Once a work request is submitted by a requestor, that requestor's supervisor should give a cursory approval of the work request and then query the work order system to see if the work being requested has already been requested, or might already exist as a work order. This small step will ensure that there are no duplicate work orders in the system. It also helps to keep from tying up maintenance with frivolous work orders.

To recap to this point, the requestor identifies a need. That requestor enters the need into the work request process and assigns the need an initial priority. The requestor's supervisor screens the request to ensure it is a legitimate need and confirms that a request or work order is not already in place for the identified issue.

The work request can now be submitted.

Now this is the critical part I mentioned earlier. The newly minted work request must be formally reviewed for approval within twenty-four hours and feedback must be given to the requestor as to the status of the request.

If we are to encourage the use of a single entry for work requests, and insist on the completion of a documented request, we have the responsibility to provide a quick and complete review of the request, in very short order. New work requests must be reviewed every day, at the same time of day. This process is often referred to as the work order approval meeting.

Approve or Decline?

Our work request process, and the work order process as a whole, has to be recognized as *the way* work gets done in our organizations. As an element of that guarantee, there must be some truth and delivery in our process.

Essentially, we are saying to the associates (remember, anyone can put in a work request), "I promise that if you'll write down your request properly, we will respond within twenty-four hours as to whether or not the work will be done."

That is a powerful promise, and one that must be kept.

Figure 6-13 shows this request turn-around and the placement

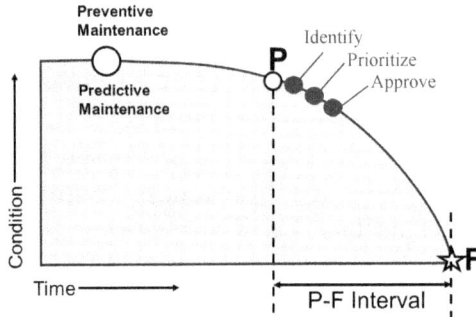

Figure 6-13: Approve nodule on the P-F curve

of the 'approve' nodule should reflect the haste with which the review process must take place.

It is during this formal work order approval meeting that the leadership decides what work truly has to get done and in what priority. In this instance, the leadership would include the production and the maintenance managers and the front line supervisors.

I would recommend that the approval for any work that requires a production asset to be taken off-line, or affects the ability of the facility to perform its service or function, be authorized by the operations manager. Any work that does not affect the service or function of the facility is to be approved by the maintenance manager. This way the person approving the work is the actual person who can grant access to the asset.

In many cases, maintenance is the default approver of all maintenance work requests. As such, maintenance essentially rubber-stamps the requests and then the maintenance department spends the next few weeks arguing over access to the machine with the production department. If you think that's silly, you'd be surprised how true it really is.

We've spoken throughout this section about a work order approval meeting, and it was suggested that this meeting be held at the same time every day. With this schedule adherence, the requestor will know within a day if the request has been approved or not, and what the additional status might be.

It bears noting, however, that the work request has to be completed correctly by the requestor. If the work request is not completed correctly, the approver should send the request back for correction. Our organizations do not have enough fluff time to track down requestors to determine exactly what they *meant* to say.

The requestor initially documented a priority on the work request. The *actual* priority is assigned to the approved work order by the approver.

I recalled a story a few sections ago (when we spoke about Root Cause Analysis) about an assignment to an oil camp in the Middle East. While I was at the camp, I attended the daily work order approval meeting. It was held every day at 7 o'clock in the morning in their main conference room. The person facilitating the work order approval meeting was the site's chief maintenance planner, but the person responsible for the outcome was the director of operations for the entire oil field.

The meeting started at the stroke of 7 o'clock. The planner brought up the very first work request from the last twenty-four hours and off we went.

At that location, every asset in that oil field complex is inspected every two weeks by an operator or a maintenance technician. Everyone is required to submit a work order request for issues they find. This includes the distant inspections of the more than one-thousand oil pumps. Interestingly, I never saw more than twenty-five work order requests a day the entire time I was at the camp.

The chief planner explained to me that, even in an enterprise as large as this oil field, if you stay on top of the work order request process, you'll never have more than twenty-five a day. That always stuck with me. And you know what? It's true.

When the first work order request came up on the screen, filled out by an operator as part of the autonomous maintenance program, the operations manager scanned it, and said one of four things:

1. OK—which meant, it is approved as a work order; keep the priority as the requestor suggested

2. 1, 2, or 3—which meant, it is approved as a work order; change the priority as I've just indicated

3. Maintenance—which meant that the maintenance manage was now to decide the status and priority

  4. Send it back—which meant, send it back to the request-
  or to complete correctly or fully

Let's key in on that very last option, "send it back." It was
explained to me that almost every job that comes in as a work
request for an oil field has to be evaluated on a few merits:

- How serious is the issue?

- Do I need to stop the asset now; is further damage or
  environmental harm likely?

- Who do we put on a helicopter and what part do they
  take with them to fix the issue?

This is something to consider when evaluating the completeness and the correctness of a work request. Is the information accurate: Do we know what the issue is, and who and what we need
to address the issue?

Remember the point made above, "Who do we put on a helicopter and what part do they take with them to fix the issue?" Keep
that in mind every time you see a work request that reads, "See
Jerry for more information."

There is much more information to cover in this chapter. This
would be a good time to summarize where we are in the maintenance
work management process. We've established these very important points in terms of identifying, prioritizing, and approving work:

- Anyone in the organization can put in a work request
  for maintenance.

- The requestor assigns an initial priority to the work
  request.

- The requestor submits the request via the formal work
  request process.

- The requestor's supervisor should review the request
  for completeness and necessity.

- The request is reviewed for approval within twenty-four hours and feedback provided.

- The approver is the leader of the agency who can grant access to the asset in question.

- The approver will approve the request based on completeness, correctness, and merit.

- The approver will return, to the requestor, any request that is not complete.

- The approver gives the work order the final priority based on other needs of the facility or location.

This summary in practice might look like Figure 6-14 as a workflow. Figure 6-14 takes into consideration emergency issues as well.

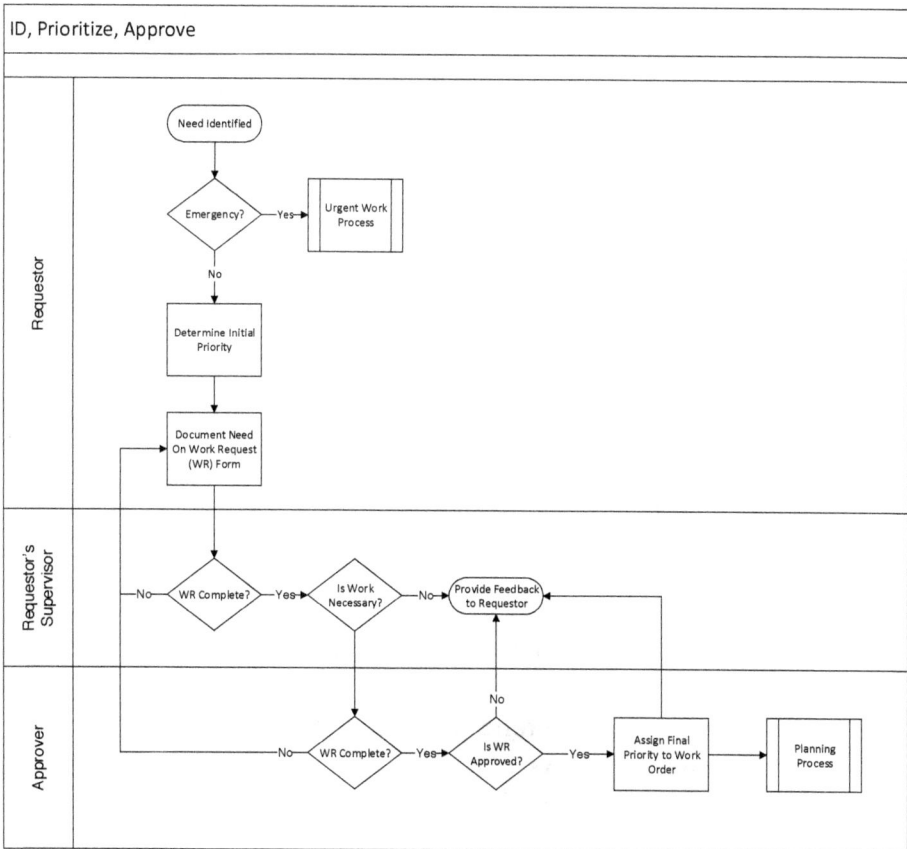

Figure 6-14: Workflow for ID, prioritize, and approve

Plan, Planning and the Planner

It should make sense that a job that is pre-planned and ready to execute would result in better quality than a job that isn't prepared in advance. The truth is, we don't get our best work in a crisis, we get our most creative work. We don't want creative work in maintenance. We really shouldn't need it.

As the title of this section suggests, a planned job requires a planner to do the planning. The unfortunate state of most maintenance organizations is that if they have a planner, that person most likely isn't doing any planning. Instead, they are likely to be shagging parts all day long. This is not the same as planning.

Where Planning Falls on the P-F Curve

Figure 6-15 indicates where, on the P-F curve, we might find the planning function. This location is where we might typically begin to think about any parts and material that might be needed to complete a specific job.

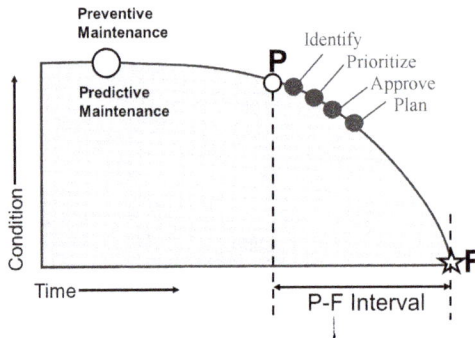

Figure 6-15: Plan nodule on the P-F curve

I mentioned a few sentences ago that the role of the planner was not to shag parts. As I just said, it is during the planning phase that we determine what parts and materials are needed. This may seem like a contradiction. Let me explain what the role of the planner really is by sharing a story that should put some real power behind this very important point.

No One Was Flying the Plane

When I was in the Air Force, I had the great privilege of being selected to attend the aircraft crash investigation course at Norton Air Force Base in California. This is a very selective program and it was an honor to attend.

During our initial studies, one of the cases we discussed was a civilian aircraft that crashed during a LaGuardia to Miami flight many years ago. The flight was one of those 'red-eye' overnight flights and the cabin was almost completely full. On the flight was a pilot from another airline. He was just trying to get home, so he was catching a 'hop' as it is called and he was allowed to sit in the cockpit on the jump seat. I'm sure the flight crew enjoyed the company.

During the flight a small light came on in the cockpit. This was a very small, inconsequential light, and not a warning light like a master caution light would be. I tell my classes that the light was probably a "check the washer fluid when you land" light.

Our crash investigation class was reading from the official NTSB (*National Transportation Safety Board*) report which included the transcript from the cockpit recordings on the black box.

In the transcript, we read that the three pilots began a conversation about the light, and what the light was for. There was the noise of rustling paper as the third pilot was apparently leafing through a technical manual, quizzing the two airline pilots of the meaning of the light. Again, it was just a light-hearted conversation about an inconsequential bulb.

Sometime during the conversation, in the wee-hours of the morning, somewhere over Florida, while everyone in the back of the plane was sleeping, one of the pilots bumped the control column. Like the cruise control in your car which disengages when you press the brake, moving the control column past a certain point in this aircraft kicks off the auto-pilot.

Unbeknownst to anyone on board, the plane began a shallow slope decent into the Florida Everglades. There was a tremendous loss of life. It was a tragic, yet avoidable accident.

I remember this case study mostly for a handwritten note at the end of the 300-page report. One of the NTSB investigators wrote, in their own hand, "Cause of the crash—no one was flying the plane."

Three pilots on board, and an auto-pilot as well. Yet no one was at the controls.

The Maintenance Planner Flies the Reliability Plane

Not to be macabre, but it is the maintenance planner in our organizations that is responsible for flying the plane. It is the planner that gets us to where we want and need to be. I make this point in the classes that I teach.

Table 5-5, in Chapter 5, laid out the roles of a planner/scheduler. For our example, these are the duties of our *pilot*. If planners gets us to where we need to be, and Table 5-5 is a listing of the roles of the people leading us to our destination, why would we allow them to be distracted from their primary responsibilities?

It's true that if a planner is off chasing parts for an emergency job, no one is going to get injured. No plane will crash. But we aren't going to get to our destination, not today anyway.

I think we can summarize the importance and the role the planner plays in our reliability efforts in one simple sentence. The planner has the singular responsibility to increase the utilization of our workforce.

Everything the planner does should be to increase the utilization of our maintenance technicians and others who are performing reliability related tasks. There should be a direct line of cause-and-effect.

The planner does *X*, which increases the utilization of the workforce by *Y*. In our work, this is what flying a plane looks like.

If you have a maintenance planner at your location, what are the typical duties of that person?

_____

_____

_____

_____

Go back over that list one more time and highlight those duties you feel increase the utilization of the maintenance workforce at your location. Be honest. Any item you highlighted would be a fundamental requirement to fly your plane, metaphorically speaking.

Now that we have a sense of the importance of the planner, and the purpose of the planner, we need to determine what a planned job really is.

What is Planned Work?

Many of the sources you and I have available to us indicate that a world-class organization works to ensure that >95% of all its maintenance work is *planned*. In order to reach anything near this high-water mark, we first need to agree on what the word *planned* means.

For a job to be considered planned, it must first be a job that we knew about. This would be a job that didn't catch us by surprise. Review Figure 6-13 again. Walk through the different nodules on the P-F curves as we've laid them out.

If you consider a singular issue, and follow the 'dots' on the P-F curve, you would agree that this would be a job that we had some pre-knowledge about. Certainly the issue was identified, prioritized and approved before it even got to the plan phase. We know about this job. In fact, we'd been talking about it several times before it hit the planner's desk.

I want to interject a very important side note here. The planner is only concerned with approved work orders. The planner never gets sidetracked with a work request. A work order, to be very direct here, is an approved work request. Do not let your planner work on assignments that haven't even been approved yet. Pilots do not go to the back of the plane and serve coffee and snacks.

Back to our description of what a planned job means. A job that is planned has to first be one that we know about. Not only for those issues identified by the plant population, but those identified through preventive and predictive maintenance. This is work that is identified back on the curve.

Additionally, a planned job is a job that we've taken steps to be prepared for. For the purposes of corrective maintenance (the result of a PM or PdM) or routine maintenance (any other work order that came in the system as a work request), being prepared for it means that we have all the parts, material, and equipment available before the job. Being prepared also means that we've listed the principle steps required to perform the work.

To sustain this process of planning a job, and to make preparation consistent, I recommend using a planner checklist. Planners should use the planner checklist for each job they walk down. Planners walk down the job to determine the detailed aspects of what is required and to note specific elements of the jobs they are planning. There are only two instances in which a planner is not required to walk down a job:

1. The planner is a master technician and has done the job in question over ten times
2. The planner was just down there earlier today

Walking down the job has the practical and political benefit of keeping the work order process out in front of the people, and keeping it visible in a positive light.

Here is just a quick synopsis to demonstrate the positive affect of that *visibility*:

- Operator detects a minor issue, puts in work order request including initial priority.
- Operator's supervisors confirms the need and checks for duplicate work orders; there are none.
- Within twenty-four hours, the work request is reviewed and approved, keeping the initial priority.
- Two days later the planner is down at the job site walking down the job.
- The operator who put in the initial work request is at the machine and talks to the planner.

Would you think that the operator in our example leaves the exchange with the planner with great faith in the work order system? I think so.

Just Follow the Checklist

So what should a planner's checklist include? It doesn't really matter what the checklist looks like, but here are some items that might be on the list:

- Work Order number

- Asset number

- Priority number

- Tools needed

- Equipment needed

- Material needed

- Parts needed

- Major procedural steps to complete the job

- Safety precautions

  - Energy isolation

  - Electronic interlocks

- Other outstanding work orders for the same asset

- Estimated time to complete the work order

- Skills required and number of people required for the work order

What other information do you think should be gathered on a planner checklist?

_____

_____

_____

_____

This information will be used to populate a job plan. The job plan is a document where we record, for perpetuity, all the aspects about a specific job.

The Job Plan

We only plan a job one time. In each instance, if we record exactly how we accomplished that job, for every part we needed, for every step we took, we would have a thorough document of that specific activity to use the next time. Documenting each specific job would result in a library that is a treasure trove of everything we've ever done at our locations.

Give this some critical thought. You deal with the same 250 component issues every single year. I just made up that number, but I am certain it is very close. Every year, you and your team are fighting and concerned with the exact same 250 components. It's not all the pumps in the plant, it's that one over there. It isn't every blower in the facility, it's the one on the roof. Not every bearing is giving you fits, it's the one on the kiln.

If, starting today, you and your teams dedicated yourselves to writing complete, thorough, yet succinct job plans for every singular component you have to deal with, within two years you'd have a complete library of everything your team would have to face in the next five years.

That is the power of the job plan.

The Value of a Job Plan—a Story

Years ago, I was in an underground mine in Canada. I was with the maintenance planner and we were going to look over the giant mining machine to do some work reviews. To get to the machine we had to travel 3,200 feet underground (yes, you read that correctly) and we had to ride in a mantrip vehicle for twenty-five miles to get to the machine.

The trip out to the machine was made even more interesting for your author, who is very much a fan of the old terra firma, by the fact that in years past, the mining crew had *nipped* an underground lake. About fifteen miles into our journey underground, we drove past an enormous pumping station. This station was pumping out of the mine the 25,000 gallons a minute that was pouring into the mine.

I want you to have the visual that the planner and I were heading to a point that was ten miles beyond the point where 25,000 gallons a minute of water was between us and the exit shaft. I don't do this every day. I was a little concerned but my host assured me everything was working as designed.

We made it to the mining machine (and back) with no issue. While we were at the machine, I asked the planner what jobs he was reviewing specifically. His answer was enlightening, as I had never heard such a response before.

The planner told me that there wasn't anything wrong with the machine. In fact, there were no open work orders, or even any pending work requests. The planner informed me that he was simply taking a day to come down to the asset to add to his list of job plans for any future work.

Huh?!

"John," he said, "There are several components on this mining machine that we don't have job plans for. I intend to write up a job plan for every one of them in case they ever do have an issue or fail so we'll have a plan of what to do."

He told me that he thought it was a good idea to add to the library of job plans while the machine was only twenty-five miles away, and not another ten miles down the tunnel.

This was a planner operating way back on the curve.

I was informed that when a machine breaks in a mine at two o'clock in the morning, it is the same as if it broke at two o'clock in the afternoon. It's dark, wet and the spare parts are hours away. My host explained that when a component failed, and the production supervisor asked the maintenance technician how long the machine would be down, the technician *now* had a *plan* to go by.

Additionally, the maintenance technician had the procedural steps on how to repair or replace the component. No creative work down in a mine. Not this one anyway.

I could only imagine the conversation that may have taken place years after I left, and that machine was more miles down that tunnel. Imagine the hydraulic pump coupling sheared and the supervisor asked the technician how long the machine would be down. "Jerry, according to the job plan, this coupling will take two hours to replace, but that's if we had the coupling with us. We don't. But I have one on the way, and although I've never replaced this coupling myself, I know exactly how to do it."

In two years' time, you could have the plan on how to address every component issue you will ever have for the next five years.

The tool the planner had with him at the mining machine that day, for our preemptive work, was a planner's checklist.

Job Plan vs. Planner Checklist

By now, you've no doubt gathered that the planner checklist is the document (it could be a device) where the planner records information while walking down a repair job. As we discussed a few paragraphs ago, the checklist asks many questions that must be answered. The information collected during the walk down, and subsequent discussions with the storeroom, maintenance technicians, and others, becomes part of the job plan.

A job plan may also include:

- Prints
- Calibration details
- Parts list
- Photos
- Copies made from the technical manuals
- Feedback form (to communicate changes with the planner)

What other items do you think should be included in a job plan?

_____
_____
_____

The most difficult detail about developing and executing a job plan process is how to file it in the CMMS so it can be retrieved in the future. Seriously, this is a major stumbling point for many organizations.

Give that some thought for a few minutes. If we start, in earnest, to develop detailed job plans for everything we do for the next two years, how do we file those electronically so we can recall them five years from now?

What do we name them? What is the protocol for creating a job plan file name? Where in the CMMS do these job plans reside? Can our CMMS even support this at all? I'd suggest that any organization starting a planned maintenance effort answer these questions in the beginning rather than after several job plans have been created.

Here is a closing thought on the planner and the chief role of the planner.

Get More Work Done

About two years after I started consulting, the day shift maintenance supervisor from the last plant I worked at called me and asked if I remembered him and if I could help him. He said, "John, you probably don't remember me, but you used to have us in your office and you'd tell us about the P-F curve, RCM, PMO, FMEA, and RCA. I was wondering if you could tell me all that stuff again, and this time I'll write it down."

My old colleague had just gotten a new job, outside that company, and he was now the director of maintenance for a very large enterprise. Leadership looks different when you are the leader, it seems.

Of course I remembered him, I said, and I'd be happy to meet with him any time to discuss these ideas and just to catch up.

When we met for lunch a few weeks later I asked him what he was trying to accomplish in his new company in his new role. He said, "I want to get my guys to work harder."

"Stop right there," I said, "You never want your guys to work harder. What you want is to get more work done." Let's be very unapologetic about this. We have got to get more work done.

A technician told me one time that he (they) couldn't get all the work orders done. I asked him when he ever thought that was going to happen. People, the work keeps coming in. We never get to the last work order. Our job, quite honestly, is to get more work done.

For my friend, we sat down and sketched out how he and his team could get more work done. You never want your guys to work *harder*. Don't believe me? Go back to Figure 2-16 and imagine that your *vision* is to have your people work harder. What would be in it for them?

At the end of our lunch, I told my friend to get a planner and start a maintenance planning process. The role of the planner is to increase the utilization of your workforce, allowing you to get more work done.

The planner *leans* out all the activities before a job, eliminating the waste that is associated with almost every maintenance activity. Or, as Gulati wrote, "Many maintenance engineers and managers consider planning to be nothing more than job estimating and work scheduling. This is not true. Planning is the key enabler in reducing waste and non productive [sic] time, thereby improving productivity of the maintenance workforce. Many organizations have started considering planning to be an important function." (*Maintenance and Reliability Best Practices,* p. 99)

One of the most involved activities the planner engages in is determining and the spare parts and materials need for a planned job.

The Storeroom

The sheer volume of work assembled to address any and all aspects of a maintenance storeroom could fill any medium-sized community's library. This is *not* a book on how to run a storeroom, but rather a section on how the storeroom contributes to work management.

As demonstrated in Figure 6-16, we have a certain amount of time in which we have to research, find, and acquire a needed part. That time is noted on the P-F curve.

Figure 6-16:
The spare parts
interval on the
P-F curve

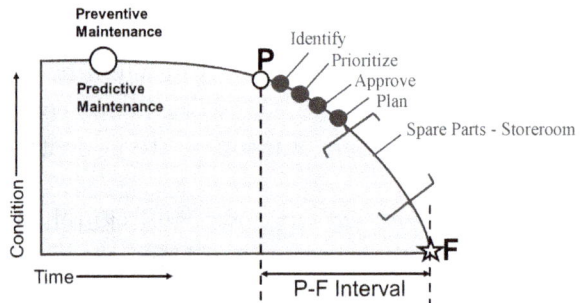

Make no mistake, the storeroom exists for the expressed and sole purpose of providing service and convenience. If a storeroom does not provide both, consistently, you'd be better off not having a storeroom at all.

Let's get our minds right to begin with on this most important subject. Figure 6-17, shows an empty storeroom.

Figure 6-17:
Empty
storeroom shelving

What's wrong with this picture? No, seriously. Show this picture to at least three different people and ask them what is wrong with this picture, which reflects empty shelves in a storeroom. In fact, it's a picture of an empty storeroom. Record their responses here:

_____

_____

_____

_____

The answer is that there is nothing wrong with this picture. Our first mistake is to believe that an effective storeroom has to stock something. That is wrong. An *effective* storeroom has to be able to *get* what we need to protect the inherent reliability of our assets. An *efficient* storeroom has to be able to get the item *when* we need it.

Neither of these definitions require a single part to be on a shelf in the storeroom, beforehand.

Our storeroom has to be effective and efficient. Our storeroom has to provide service and convenience. Our storeroom doesn't have to stock anything to deliver on all mandates. I'm willing to bet that nearly everyone reading this disagrees with me. That's perfect.

Service and Convenience—an Example

There are any number of auto parts stores around the U.S. Most of them are laid out and operated in a similar fashion; in fact, some may even be virtual and web-based.

When you walk into a brick-and-mortar auto parts store, the first thing you might notice is that it is well lit, usually clean and organized, and seemingly well stocked. What is out in front of the service counter is all of the items the store management wants you, the customer, to help yourselves with. If you want some mud flaps, a license plate holder, or some wind shield wipers; help yourselves.

But if you want a real car part, you have to go to the service counter and ask for it.

The counter attendant is going to ask what you need, and you'll either present them with the part you need, or describe the part you need. When describing, you will need to provide the make and model of the vehicle, and as much detail on the part as needed for the attendant to understand what you need. At no point is the attendant going to invite you back to find the part yourself. Never.

Once you've communicated the part that you need, the attendant is going to tap on their computer and say one of three things:

1. I'll be right back.

2. We don't carry one, but I can get one from the next town over. If you'd like to come back this afternoon we'll have it for you. Otherwise, we can ship it to you for free. What would you like to do?

3. We don't carry one, but I can get one from the closest major town. If you'd like to come back tomorrow we'll have it for you. Otherwise, we can ship it to you for free. What would you like to do?

Now that is service and convenience! It's also effective and efficient.

Effective First, Then Efficient

Many leaders in today's organizations feel that their storerooms are not efficient. Again, efficient is a word that you have to challenge people on as to what they actually mean. We need to first be concerned with the effectiveness of our storerooms before we trouble ourselves with the efficiency.

When I'm asked to help a storeroom convert to barcoding, I first check to make sure they have written procedures and are following those for all their storeroom interactions. The truth is, if you can't master pencil and paper, a bar code scanner isn't going to make you more effective.

The goal of the storeroom is to provide the right part, at the right time, in the right quantity. This is the fundamental guiding principle. Let's face it. We are not going to have a storeroom full of empty shelves. Neither should we have a storeroom full of stocked shelves that are stocked with the wrong parts.

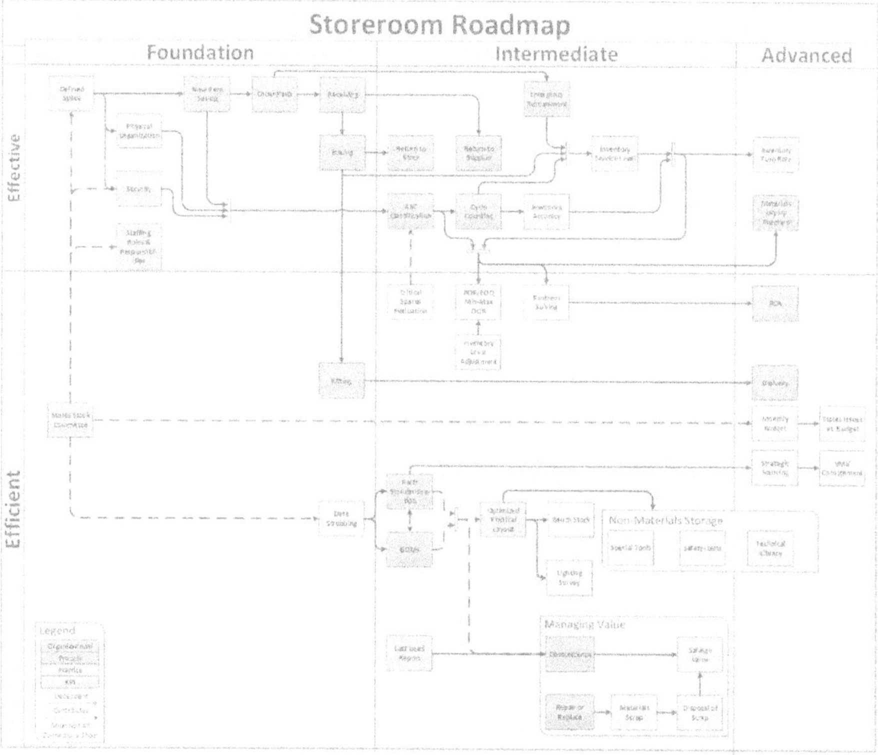

Figure 6-18: The storeroom road map

Figure 6-18 is a road map that a colleague and I developed many years ago, and it is one of the tools I use to determine where a company is on its journey to world-class. It is quite literally a road map.

As you can see in Figure 6-18, there are three columns and
t   w   o
rows. I always start consulting on storeroom operations by ensuring
the client has the Foundational-Effective section nailed down.

There are eight blocks in the Foundational-Effective section:

1. Defined space—What is the jurisdictional area of the
   storeroom?

2. Physical organization—Who is responsible for the store room?

3. Staffing, R&R—What do the storeroom associates do?

4. Security—How is the inventory guarded and protected?

5. New item set-up—How does the organization identify what
   parts to stock?

6. Ordering—How are stocked and non-stocked items ordered?

7. Receiving—How are items received in the storeroom and
   reconciled with the purchase order?

8. Issuing—How are all these items issued out and accounted
   for?

It is these eight very foundational elements that ensure the
storeroom will have the right part at the right time and in the right
quantity.

Right Part, Right Time, Right Quantity

The storeroom as an entity should only stock what it is instructed
to stock and in the quantity requested. To round this essential point
out, the particular stock has to be made available at the required
time. This after all is the goal. But why is it the goal?

Figure 4-6 is a representation of Mean-Time Between Failure.
Marry that image with the overall responsibility of the maintenance organization, *to protect the inherent reliability of the asset,*
and you might begin to see that the *right part* is essential to increasing MTBF.

Where does the right part come from?

The right part initially comes from engineering, which, if following standardization practices, has designed the asset or the mod-

ifications with the right part to begin with. All this part data is communicated on the *Bill of Materials*. We spent time earlier discussing the BOMs.

The storeroom will have the right part if the associates are told what the right part is. Who tells them this?

The maintenance department will tell the storeroom what to stock and in what quantity. Subsequently, the maintenance department is also responsible for telling the storeroom what to get rid of (obsolete parts). The maintenance department knows what parts are to be stocked based on the BOMs that the engineer has delivered. The actual person that formally requests the part to be stocked or removed is the maintenance planner. It is part of flying the plane.

Figure 4-5 is a representation of Mean Time To Repair. One of the most significant elements that make up MTTR is the whole issue of 'parts' and their availability. Having the part at the *right time* and in the *right quantity* is essential to decreasing MTTR.

The storeroom will have a better chance of having the right part at the right time and in the right quantity the further we operate back on the P-F curve.

At any time in these few paragraphs have we discussed anything that indicates that we absolutely have to have a stocked storeroom? The answer is no. We agreed that a storeroom should be efficient and effective plus provide service and convenience. There has been no discussion to suggest that a fully stocked storeroom is necessary.

We also just detailed that the right part, at the right time, and in the right quantity leads to increased MTBF and decreased MTTR. These are notable reliability metrics. Still nothing to indicate, much less demand, a shelf full of parts.

It was just recently mentioned that the storeroom has a very good chance to get the needed parts, the further we can identify needs back on the P-F curve.

Why is it that we stock stuff anyway?

What If?

We stock stuff because we don't really know what's going to happen. Let's face it, we'll always have emergencies as long as we have people driving around our facilities on fork trucks. They're bound to run into something. We increase the odds of that by painting stuff yellow so they can see it through the fog of the night to run right smack into it.

Our storerooms become the manifestation of our best guess. Our stock becomes those items that, based on our professional opinions, we'd be nuts to try and run the plant without. Our storeroom inventory is the result of our expert advice.

Are we so lacking in technical or anecdotal knowledge that we have to stock a mountain of stuff just in case? I think we can do better.

I'm going to show you a *stock or don't stock* decision tree in a few paragraphs, but I'd like you to give this a try. Answer this question yourself, or better yet, if able, gather a few people and ask, "If we had to fill an empty storeroom with stocked items, what criteria would we use to determine if we should stock it or not-stock it?"

Let that question marinate for a short time and then record

_____

_____

_____

_____

some thoughts here:

I've thought about this for almost three decades and I finally did something about it. Figure 6-19 is what I came up with as a decision tree to answer this question. A side note here: I'd recommend that we always have a decision tree or a flow diagram to follow. We should not make key maintenance and reliability decisions based on bias or personal gut-feel. See the section of Critical Thinking if you think everyone approaches an issue with the same mindset.

The very first section of the Stock, Don't Stock Decision Tree is concerned with our knowledge of the function of the component. Do we have any idea of what this part does and why it matters? That is basic Reliability Centered Maintenance.

The second level of questioning asks whether the part exists on a BOM for an asset in the facility. If not, why are we even stocking it? It doesn't belong in a high-performing storeroom.

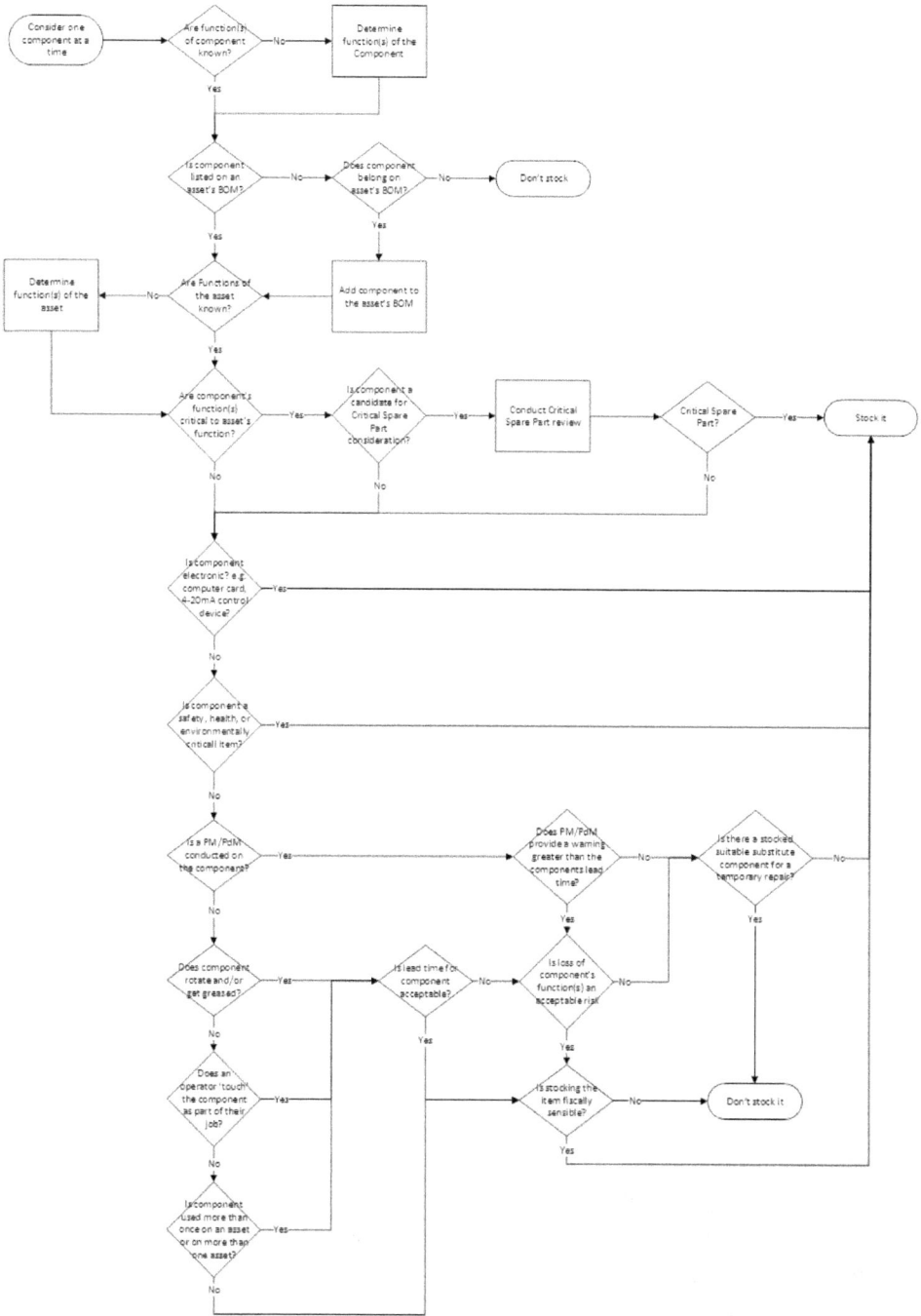

Figure 6-19: Stock, don't stock decision tree

If a part for consideration makes it past those two hurdles, do we know what function it serves on the asset it is assigned to? This helps us determine the degree of criticality we need to consider for this part.

The real meat and potatoes of stocking an item or not comes in the next series of decisions. This is purposefully set up to really make us think of what the part does, and how we know where it is on the P-F curve. These are hard questions and need straight answers.

Establishing a world-class storeroom and stocking it to perform as such is no pedestrian affair. This is serious work for serious people.

We are limited as to what we can stock. If not limited by a cap restriction on inventory value, we're restricted by volume. From the storerooms that I've worked with, it is clear to me that we've used up all the square footage, and now we're working on the cubic footage.

What limits us?

A Historic Example

I've made this comparison many times and at one time reduced it to print. I think this is a fun, historical example which will bring to mind that there is a difference between what we want and need. There is a significant difference between what we think our capacity is and what it actually is.

I live in Kansas City, Kansas in the Westport District. If you're not from that area you most likely aren't familiar with the districts of KC. But, like most cities, Kansas City (Kansas and Missouri) is divided into areas or districts, ostensibly a by-product of the communities that came together to form the metropolis.

Westport is famous among history buffs. This is the area where the Santa Fe and Oregon Trail wagon trains provisioned-up for their journeys west. It was one last stop in civilization to get those critical, often life-preserving supplies, before heading out into the great unknown.

Imagine those times, those hopeful events, and all the planning and preparation that went into a logistic execution that quite honestly, your life depended on. Things really haven't changed that much in 150 years.

What do you take, what can you take, what's your limit? How do you preserve perishables while you're traveling? Are there other provision points along the way? What do I have or need that is repairable, how do I repair it, and what do I need in order to repair it? How about defensive equipment and supplies? You were limited by what your team of horses could pull. Each wagon or grouping of wagons had to be a bit self-sufficient, but the enterprise, as a whole, had to be creative in sharing and resourcing amongst themselves to get everyone safely to their destinations. You didn't want to be a burden on the wagon train, but you couldn't take everything. Did you need it or just want it?

Fast forward 150 years, and in many ways, we're still having the same discussions. Wagon trains, and indeed the development of the West, were successful in part because people learned and improved the process as they repeated the journey. The steam locomotive came later, but the earlier pioneers had to count on the long and hard route.

The discussion we need to have today is one that has been perplexing us since those days, and quite honestly, for much longer in our history. How do we know what to take, or more specific to our discussion, how can we be sure what to stock or not stock? In some small way, each of us is in a small collection of people trying to survive and thrive in our environment. Some operate in larger cities with many resources, others in smaller communities where they are forced to operate where resources are light.

In an attempt to answer this stocking question, I generated a robust decision tree. I have, over time, come to assemble what I feel to be a very effective tool to arrive at a measured and well-vetted decision. It's critical at this point that the reader understand that the decision to stock or not to stock an item should not be an emotional, or biased decision (or worse, made on a whim). It is one that should be measured, pragmatic and applies a standard and repeatable logic. Imagine a team of four horses pulling the weight of your storeroom along the entirety of the Santa Fe Trail. You simply don't have the capacity.

This is a short section on the storeroom; short compared to the collection of knowledge that exists in our industry. The attempt here was not to make everyone an expert on all the processes of an effective and efficient storeroom, but rather to answer this simple question in closing.

If our storeroom is to have the right part, at the right time, in the right quantity to provide service and convenience, what do we put on the shelves in Figure 6-17?

This whole time we've been sliding down the P-F curve. Once the parts are available, it's time to schedule the job.

Schedule, Scheduling and the Scheduler

Planning *leans* out the work before a job is executed. Scheduling *leans* out the empty space between the jobs.

Scheduling is the activation of planned work, or for the examples we've been discussing, scheduling is the activation of the job plans we've created. We schedule planned jobs, not unplanned work. Recall that planned jobs are jobs that we are prepared to execute. We don't schedule a job until we are ready with everything we need to execute the work. Many reading this text will agree that it is very easy to be 'burned' by not having a part available at the time needed to actually perform a job as scheduled.

I myself have seen a maintenance technician perform a PM on a drive belt, finding it in need of replacement, and cutting the belt off before going to the storeroom to get the replacement. I'm sure you can guess how that turned out.

The work management act of *schedule* or *scheduling* work falls somewhere on the P-F curve as shown in Figure 6-20.

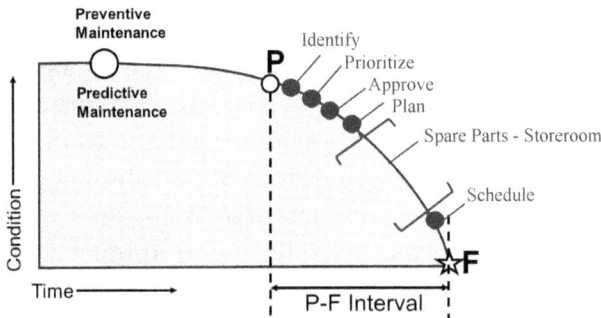

Figure 6-20: Schedule nodule on the P-F curve

The ambiguous phrase " ...falls somewhere on the P-F curve as shown... " simply reflects that the job will be scheduled for execution sometime after the parts are confirmed to be present and available.

Scheduling has become, perhaps it has always been, undoubtedly the hardest executable segment of maintenance work management. This is simply because we haven't figured out how to play together in the sand box with production.

Production Won't Give Us the Machine to Work On

Deming spoke of *constancy of purpose*, admittedly a phrase that I was not familiar with until I became a student of equipment reliability. Where Deming was arguably discussing a 'business' responsibility to become competitive for the good of the shareholders, I like to co-op this term to suggest that we all need to be working together for the same reason, or purpose.

Consider the diametrically opposed purposes that production and maintenance have, although they are working in the same facility.

We've mentioned before that everyone in the facility owes, in a manner of speaking, their living to the performance of the assets. It is the assets that make the income, from which we pay the expenses, which is your paycheck. To accomplish this *purpose* the asset needs to be running. At least it needs to run longer than it doesn't run.

In order for production to do what it is asked to do, for the intended reason of having production, the machine needs to be running. Operations better fulfills its purpose if the asset is running.

Maintenance, on the other hand, more often than not, needs the equipment to be in a state of 'not running' in order to complete what they feel to be their secondary *purpose*—fixing the equipment. It is ironic to think that maintenance can keep the equipment running (primary purpose) by stopping it to fix or inspect it (secondary purpose).

Production needs the equipment to run. Maintenance needs the equipment to stop. There are twenty-four hours in each day. Now what?

Look back at that simple change model in Figure 2-16. I want you to imagine this conversation with the production leadership at your location. "Our vision in maintenance is that you stop the equipment more often, and as a result, you will make more production."

What's in it for them?

When we are asking production to stop the equipment for us to inspect or to complete programmed preventive maintenance, isn't the implication that they will run better and longer afterwards?

Have we ever completed a PM on a machine and we can't get the machine to start up again? Have we ever completed a PM and the same component broke again just a few days or weeks later? Take a look at Figure 6-5. Maybe a component we changed out or a modification we made a couple of weeks ago didn't quite work out.

Let's face it. Sometimes our maintenance isn't so good. Our 'street-cred' is low. Is it any wonder that operations isn't more excited about giving maintenance the machine more often? They never know what condition it's going to come back to them in.

The good news is that we are correcting all that. We know how to improve the quality and results of our preventive maintenance through the PMO process. We know that a dedicated PM team helps to build a high performing proactive maintenance program. We also have a bead on the importance of designing, building, and installing an asset correctly to ensure a machine with high inherent reliability. And recently we agreed on having the right part, at the right time, in the right quantity.

We in maintenance have done a lot here. Can we just get operations to play ball with us?

When to Schedule Maintenance

I promise you that this will be the most math intensive section of this entire workbook. In fact, you will need a calculator, pencil, and paper—or just write in the margins. What we are going to be exploring is a method that you can use very soon to make a compelling argument for a predicable maintenance schedule. This section alone is worth the price of this book!

Ramesh Gulati's great work, *Maintenance and Reliability Best Practices*, includes a section on calculating reliability of an asset and continues with calculating the reliability of a line of assets that are set up in series or parallel. We are not going to go into all of that, but rather, we are going to look at a singular reliability number and determine the maintenance schedule for that entity.

The Back Story

The very first time I did this process, I was at a corrugated box company out west. I had somehow gotten myself into a discussion, an argument really, between the operations manager and the maintenance manager.

The operations manager wanted to run a particular machine for twenty-four continuous hours. The maintenance manager was telling him that he couldn't do that. Now, I know for you and me, running a machine for twenty-four hours is nothing compared to almost everything in your plant. But this was the scenario taking place before me, and I was merely a fly on the wall.

The operations manager looked at me and said, "John, the Mean Time Between Failure for this machine is twenty-six hours. That means that I can run it for twenty-six hours without a breakdown, right?"

I told him that wasn't exactly what Mean Time Between Failure meant, and by the way, twenty-six hours was horrible!

I showed him the formula in Fig. 6-21 which is slightly modified from Gulati. (*Maintenance and Reliability Best Practices*, p. 163)

Figure 6-21:
Reliability calculation

$$e^{-\lambda(t)}$$

As Gulati explains:

- e is the natural log; it is a constant value, which is approximately 2.71828

- λ is the Greek letter lambda; in this formula it represents the inverse of the MTBF

- (t) is the time that the asset is intended to run, based on the same units as the MTBF (hours, minutes, etc.)

What I sketched out for the operations manager was something similar to Table 6-4.

Table 6-4: The client's reliability calculations

Intended run hours	λ	Resulting reliability
4	8	.9 or 9 %
2	8	6or 8%
8	8	3or 9%

I told the production leader that with a MTBF of twenty-six hours, he had about a 39% chance of getting through a twenty-four hour straight production run without a hiccup. In fact, I said if I was going into a production run, I'd want a reliability of greater than 85%.

Test your proficiency with the formula in Figure 6-21. I'll provide the scenario:

You're A-1 packaging machine is a stand-alone asset and is used sparingly during off-peak production and more often during peak production. You and your team are confident in your calculation of the MTBF of this equipment, and the number you've arrived at is 560 hours.

The MTBF is 560 hours; what is the inverse, or lambda?

_____

_____

The production team wants to run this packaging machine for thirty straight days, around the clock. This is the peak period.

What is the resulting reliability of that scenario?

_____

_____

I realize that I'm using reliability, probability, and chance interchangeably, but the synonym works for this scheduling discussion.

You have calculated the lambda, given the MTBF figure, and you have calculated the reliability figure using the formula in Figure 6-19. Now calculate what continuous run time would be permissible to have a resulting reliability (probability) of getting through that run time of 85%. What is your answer?

_____

_____

My answers:

Lambda = 0.0018 (the inverse of MTBF)

The probability of getting through 30 days = 27%

For a probability of 85% = run for 90 hours at a time, or 3.75 days.

Now hold on, this is the interesting part and it is how this all ties into the schedule. With the numbers as just calculated (use mine or whatever your values are) Figure 6-22 would be the proposed schedule approach to this particular machine.

90 hrs. running	1 hr. maint.	90 hrs. running	1.5 hrs. maint.	90 hrs. running	1 hr. maint.	90 hrs. running	2 hrs. maint.	90 hrs. running

Figure 6-22: Proposed run schedule for the A-1 packing machine

Based on the historical run record of the A-1 packing machine, our numbers tell us that we can run for ninety continuous hours with a *probability* of success (no unplanned downtime) of 85%, or perhaps greater. This is a very loose, yet I believe a practical way to use historical data to determine how often we should perform maintenance on an asset.

The maintenance times shown in Figure 6-22 are just examples of downtime periods. The numbers you would use would be the real numbers. The question now becomes, "what do we do during those down time sessions?" I suggest we perform the required PMs and corrective planned maintenance. We really want to do what a colleague of mine said many years ago, "just do something awesome."

I do want to come full circle on this approach and go back to the discussion of equipment excellence. During these periods of downtime that are programmed into our equipment run cycles, we are going to be addressing the very components that have caused our less than ideal MTBF. In fact, there ought to be an army of reliability engineers working to increase our MTBF from 560 hours to 700 hours to 1,000 hours, and so on. That is the job of a reliability engineer!

If you ran every asset in your facility with this unbiased and truthful scheduling scheme, you'd have programmed and predictable windows of opportunity to inspect and repair production equipment. Production would get assets that don't fail in the middle of a production run.

For a practical example, consider this very short story. I used something very similar to this formula and approach as the general foreman for maintenance at a Midwest steel mill. I convinced my boss, the person who ran the steel mill melt shop, and his boss, the Vice President for Operations for the entire company, to let me try this scheduled maintenance approach. In the first year, we made and sold $200 million more in product than was forecasted. How? Except for the scheduled downtime, we never stopped. By the way, you read that profit number correctly—$200,000,000.

We know that a programmed and predictable schedule can be enforced, but what about those one-off scheduling instances?

Using the P-F Curve to Win a Scheduling Argument

The section we just finished covered an overall approach to scheduling. However, during our routine preventive and predictive or autonomous maintenance, we are likely to discover a component that needs repair or replacement. Instances such as these are what cause much of our scheduling grief.

Let's consider as an example that our PM/PdM crew picks up a high vibration during our condition-based monitoring on a 50-Hp motor. As a result, maintenance has walked through all the steps outlined for high vibration on a motor. The bearings were greased and the motor was checked for secure mounting and proper alignment. All the steps proved ineffective, and the only step left was to replace the large motor.

As the Condition Based Maintenance plan had been agreed to by operations and maintenance, it would seem a simple task to schedule the four hour motor replacement. However, production seems to be reluctant in providing the down time window to a replace a motor that is quite literally 'still running.' Where is the disconnect in the partnership with operations?

What do we do? What would you do?

Figure 6-23 is the P-F curve argument.

This sounds simple to you and me. However, reflecting on Figure 6-23, please notice that the entire time maintenance is addressing the issue and discussing it outside the maintenance organization, the motor is 'sliding down' the P-F curve.

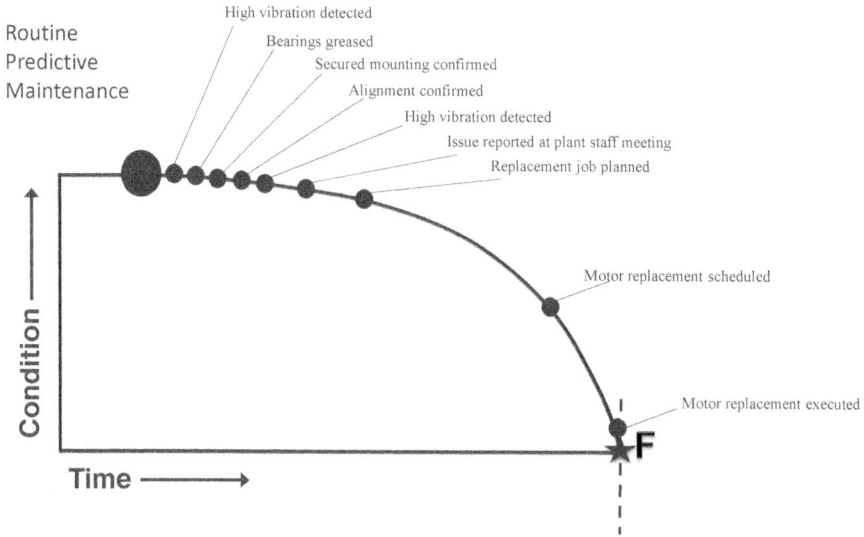

Figure 6-23: P-F curve argument for time on the asset

My bigger point is this. If maintenance detects an issue or an associate identifies a concern, we have to act. Maintenance is responsible for having its act together. Our job, before we are turning wrenches, is to provide technical advice. If production doesn't take that advice, they can't report they are down for maintenance; no, they're down because they made a bad business decision.

Our maintenance advice has got to be rock solid. We have to detect things as early as we can on the P-F curve, and provide exceptional direction based on facts. We, in maintenance, have to provide evidence of our discovery and work. Production knows they matter more and make more product if the equipment is running. We need to convince them of another idea.

Someone once told me that the ability for maintenance to stick to a maintenance schedule is dependent upon production's ability to stick to a production schedule. I think we are all in this together.

Now it's time to get the job done!

Execute Maintenance

Well, you don't execute maintenance—I'm pretty sure that would be against the law. What we execute is the maintenance plan as shown in Figure 6-24.

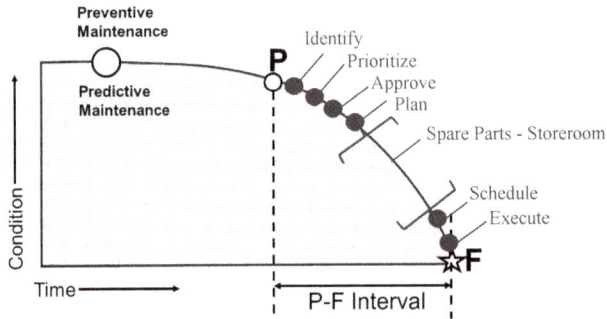

Figure 6-24: Execute nodule on the P-F curve

In both Figure 6-23 and here in Figure 6-24, you'll note that the point where we *execute* maintenance is very close to the point of function failure. This is done for a reason that is fiscally sensible.

If you take, for example, the 50-Hp motor we used in the example for Figure 6-23, consider that a 50-Hp motor can cost $3,500. For that particular motor, we want to get $3,499 worth of value before we pull it out of service and send it out for rewind or rebuild (if that is even possible).

This approach to scheduling and executing should not be a surprise to anyone in our organization. In the strategy that we develop, we have to include this routine practice of executing maintenance right before a functional failure. In fact, our philosophical approach should be that, together with production, we maintain all assets and components to get full value for the money.

Even with an agreeable philosophy, if we don't execute the maintenance plan at the execute point in Figure 6-24, we are going to break down while in production.

If your organization runs 24-7, blowing past the scheduled execution date would guarantee a 67% chance of failing on a shift other than day shift. In fact, day shift is when your most senior and knowledgeable maintenance technicians are scheduled (I purposely did not say your most 'capable' technicians). Day shift is also when our contractors, vendors, and suppliers are open for business. Day shift is when other businesses will answer their phones. We need these agencies to be available for contingency operations, if needed.

Every single person reading this workbook knows this next point to be true. We just don't do anything about it.

Almost all maintenance should be executed on day shift, Monday through Friday. I say 'almost' all maintenance to account for extended downtime needs such as rebuilds and major repairs. All maintenance would include all planned maintenance, with the notable exceptions just listed.

Circle the response that most closely reflects your maintenance execution methodology:

When convenient for production
or
Whenever necessary, regardless of convenience

Most would circle 'when convenient for production.' There is nothing wrong with that, and it may be that it is not wrong or in error in any way. It does make sense that maintenance is conducted when the equipment is *naturally* available. These are normal scheduled non-production periods that we'd be fools not to take advantage of.

If your facility or your assets run 24-7, what's the difference between 1:30 p.m. on Tuesday and 8:00 a.m. on Saturday? A production manager once asked me why I was complaining that he never gave us the machines to work on. His message (and this is absolutely true), "You have Easter, Memorial Day, the 4th of July, Labor Day, Thanksgiving, Christmas and New Year's Eve and day. Those are three-day weekends. What are you complaining about?"

Raise your hand, metaphorically, if you have had significant maintenance scheduled for the weekend, only to be informed on Friday afternoon that production was going to run into the weekend. I had a client who had a repair technician from Europe on site for $25,000 a day. Production decided to run on Saturday and Sunday. The technician and his company were happy to have him enjoy the area and local sites for seven days until the next opportunity, which was the following weekend. You cannot make this stuff up.

Have you ever contracted for a heavy lift crane to be on site only to have the crane operators sit there with nothing to do while production ran? This is an example of how the entire scheduling-executing model is broken.

But we can fix it.

The Proper Way to Execute Planned and Scheduled Maintenance

During our strategy development we are going to set into motion the manner in which we hope to execute planned and scheduled maintenance. We are going to state that we hope to execute all planned maintenance, which includes PMs and corrective, and routine maintenance during normal business working hours. Normal business working hours are for us, our contractors, and our vendors.

This should be the model routine we incorporate into all our planned maintenance:

1. The operator is aware of scheduled maintenance; machine is still running.
2. The maintenance technician arrives on time, with materials and equipment.
3. The maintenance technician and operator discuss other issues that might be present.
4. If there are other issues, the maintenance technician will call their supervisor to the machine.
5. The maintenance technician asks the operator to keep the equipment running while the maintenance technician walks around the asset performing a look, listen, and feel inspection.
6. The maintenance technician will alert the operator and the maintenance technician's supervisor if other concerns are detected.
7. The maintenance technician instructs the operator to clear out their machine, and shut it down in a controlled manner.
8. The maintenance technician locks out the machine.
9. The maintenance work is executed.
10. The maintenance technician removes the lock out.
11. The operator restarts the machine while the maintenance technician is still present.
12. The operator runs five 'good' parts (or the equivalent).

13. The operator confirms proper operation.

14. The operator signs off on a survey on the back of the work order indicating their satisfaction with the maintenance that was accomplished.

15. The maintenance technician signs off on the survey on the back of the work order indicating their level of satisfaction on how ready the asset and the operator were for the scheduled work.

16. The maintenance technician completes the work order.

17. The job is done.

What are some additional steps you use in this process, or that you'd like to see:

_____

_____

_____

_____

Scheduled maintenance is a 'dance' of sorts. There are at least two parties involved, most commonly maintenance and production. Each has to know their own moves and how each person contributes to the success of the job.

Take a highlighter or a pen and annotate on the list just created those individual line-items that are not accomplished when we do all our maintenance on a weekend.

During routine maintenance at your location, what is the role and responsibility of the operator?

Record your thoughts here:

_____

_____

_____

_____

Tying it Together

We've done some really good work over these last few pages. We are now going to use the points we've previously made to build a really good execution strategy.

Look back at Figure 6-22 and Figure 6-23. Figure 6-22 provides us a glimpse into a predictable scheduling model that reflects what an asset has historically delivered. It gives us a very good idea of how long a machine can run uninterrupted. Figure 6-23 is a pictorial representation of a specific concern.

Although presented as two independent thoughts, these two ideas actually merge to form the genesis of our execution approach.

Here is the scenario:

Ajax Printing runs its Heidelberg press using a predictable maintenance schedule based on the historic MTBF rates recorded over the last three years. The historic MTBF on the press is 2,000 hours and the reliability engineer has identified the components that have this machine throttled back. Projects are being considered that would increase this MTBF to 3,500 hours.

Until then, the company has elected to run this press for 336 straight hours and then provide one shift of maintenance downtime. The operators have their duties to perform during the maintenance down shift, and maintenance is charged with getting everything done during this time frame.

The run schedule, until further notice for the Heidelberg press, is shown in Table 6-5.

Table 6-5: Ajax Printing's Heidelberg run schedule (partial)

Sunday	Monday	Tuesday	Wednesday	Thursday	Friday	Saturday
Run	Run	Down Day shift	Run	Run	Run	Run
Run	Run	Run	Run	Run	Run	Run
Run	Down Afternoon shift	Run	Run	Run	Run	Run
Run	Run	Run	Run	Run	Run	Run
Down Midnight shift (going into Monday	Run	Run	Run	Run	Run	Run

Given all that we've been speaking about these past few pages, circle the date above that you, as a leader in your organization (maintenance or operations) would recommend for a four hour 50-Hp motor change out. Explain your reasoning here:

Yours, Then Mine, Then Yours Again

Without doubt, the most neglected or abused element of *execution* is handing over the equipment to maintenance and then maintenance giving it back to production once the job is completed. For that reason, I am reprinting the list we made just recently:

1. The operator is aware of scheduled maintenance; the machine is still running.

2. The maintenance technician arrives on time, with materials and equipment.

3. The maintenance technician and operator discuss other issues that might be present.

4. If there are other issues, the maintenance technician will call their supervisor to the machine.

5. The maintenance technician asks the operator to keep the equipment running while the maintenance technician walks around the asset performing a look, listen, and feel inspection.

6. The maintenance technician will alert the operator and the maintenance technician's supervisor if other concerns are detected.

7. The maintenance technician instructs the operator to clear out the machine, and shut it down in a controlled manner.

8. The maintenance technician locks out the machine.

9. The maintenance work is executed.

10. The maintenance technician removes the lock out.

11. The operator restarts the machine while the maintenance technician is still present.

12. The operator runs five 'good' parts (or the equivalent).

13. The operator confirms proper operation.

14. The operator signs off on a survey on the back of the work order indicating their satisfaction with the maintenance that was accomplished.

15. The maintenance technician signs off on the survey on the back of the work order indicating a level of satisfaction on how ready the asset and the operator were for the scheduled work.

16. The maintenance technician completes the work order.

17. The job is done.

The *exchange* process is somewhat likened to the passing of a baton in a relay race. At any given time one person is responsible for the baton. But an integral part of the race is the careful and well executed *passing* of the baton. This is the 'hand off.'

Review this list again. At one point the operator has the machine, or baton. Next it is in the hands of the maintenance technician. In the closing part of the race the machine (baton) is handed back to the operator. It is a dance, a race, a rhythmic arrangement to ensure everyone is doing their part.

Has it ever happened in your organization that maintenance is complete with a maintenance assignment, and the operator is nowhere to be found? Has a maintenance technician ever gone home and left their lockout on a motor disconnect?

How is that even possible? How many times a day do we execute this hand off?

The two processes, operations handing the baton to maintenance and then back again, have been referred to as *turnover* and *handover*. Although these two critical processes are not often formally mapped out in a flow diagram, they are very important and often prove to be the difference between a good exchange and a fumbling of the 'baton.'

Yours, Then Mine—The Turnover
From the point made just earlier:

1. The operator is aware of scheduled maintenance; machine is still running.

2. The maintenance technician arrives on time, with materials and equipment.

3. The maintenance technician and operator discuss other issues that might be present.

4. If there are other issues, the maintenance technician will call their supervisor to the machine.

5. The maintenance technician asks the operator to keep the equipment running while the maintenance technician walks around the asset performing a look, listen, and feel inspection.

6. The maintenance technician will alert the operator and the maintenance technician's supervisor if other concerns are detected.

7. The maintenance technician instructs the operator to clear out the machine, and shut it down in a controlled manner.

8. The maintenance technician locks out the machine.

This is the classic operations to maintenance turnover.

The measure of our collective success with turning over equipment from operations to maintenance is made during the survey completion after the job is finished.

Recall that everything done in maintenance is a process, and as such, everything can be mapped out and associated with a RACI chart and process guide.

If we consider the steps just reprinted, and add in a few more logical flow ideas, we might arrive at a workflow similar to that shown in Figure 6-25 on the following page.

Figure 6-25 takes into account those instances where the asset is not available when the maintenance technician arrives, or cannot be made available. The idea of a daily scheduling meeting is introduced in this workflow as well.

Many organizations have compressed the number of meetings they have into a singular meeting, often called a YTT, or *Yesterday-Today-and-Tomorrow* meeting.

The first part of the meeting, the Yesterday portion, revolves around reviewing for approval the work requests that have been submitted over the last twenty-four hours and the results of yes-

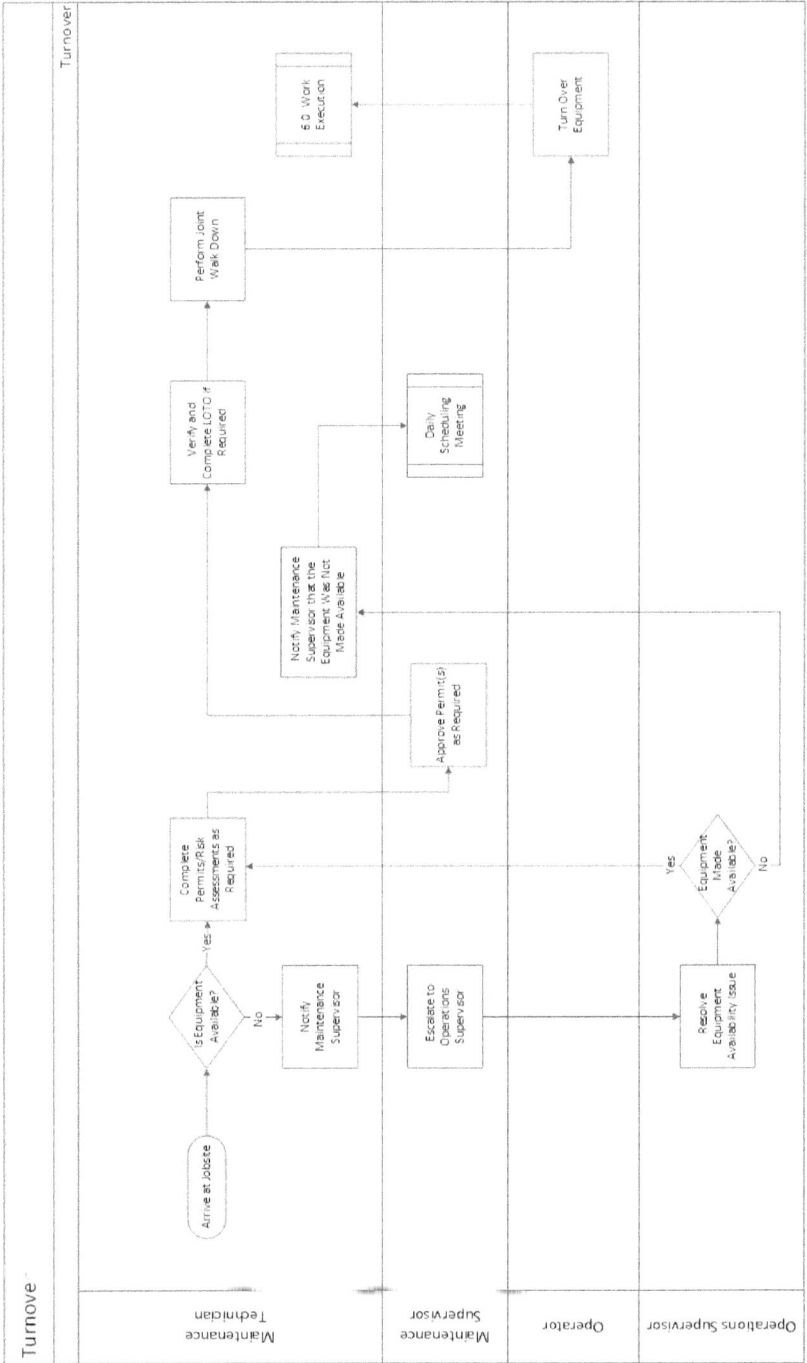

Figure 6-25: Turnover workflow

terday's maintenance. The Today and Tomorrow segments are what we are speaking of in this example. This is essentially the daily scheduling meeting. It forces us to address the current scheduled work situation.

Then Yours Again—The Handover

Reexamining from our original list, the steps that concern us in this process are the latter steps.

10. The maintenance technician removes the lock out.
11. The operator restarts the machine while the maintenance technician is still present.
12. The operator runs five 'good' parts (or the equivalent).
13. The operator confirms proper operation.

The measurement of the success we have in returning the equipment to service or handing it back from maintenance to operations is made during the survey completion after the job is finished.

These surveys are important as they provide the 'vehicle' by which the operator and maintenance technician can each give feedback to one another in the hopes of conditions improving. There has to be some form of measure on how effectively (and efficiently) we pass the baton among each other.

Since we're talking about surveys, let's clear up a few points. If we ask anyone to fill out a survey, that person has every right in the world to believe that we are going to use the survey results to make things better. There is no quicker way to kill an improvement effort than by surveying everyone and then doing nothing about it.

We should take the survey notes annotated on the back of each work order and work collectively to eliminate causes (which would include associates) that keep us from having high marks.

Back to the handover.

Like we saw in the turnover, the handover steps can also be added to in an effort to arrive at a complete workflow. This process, as mentioned earlier, may be referred to as *return to service*.

Figure 6-26 might look like the process flow we develop.

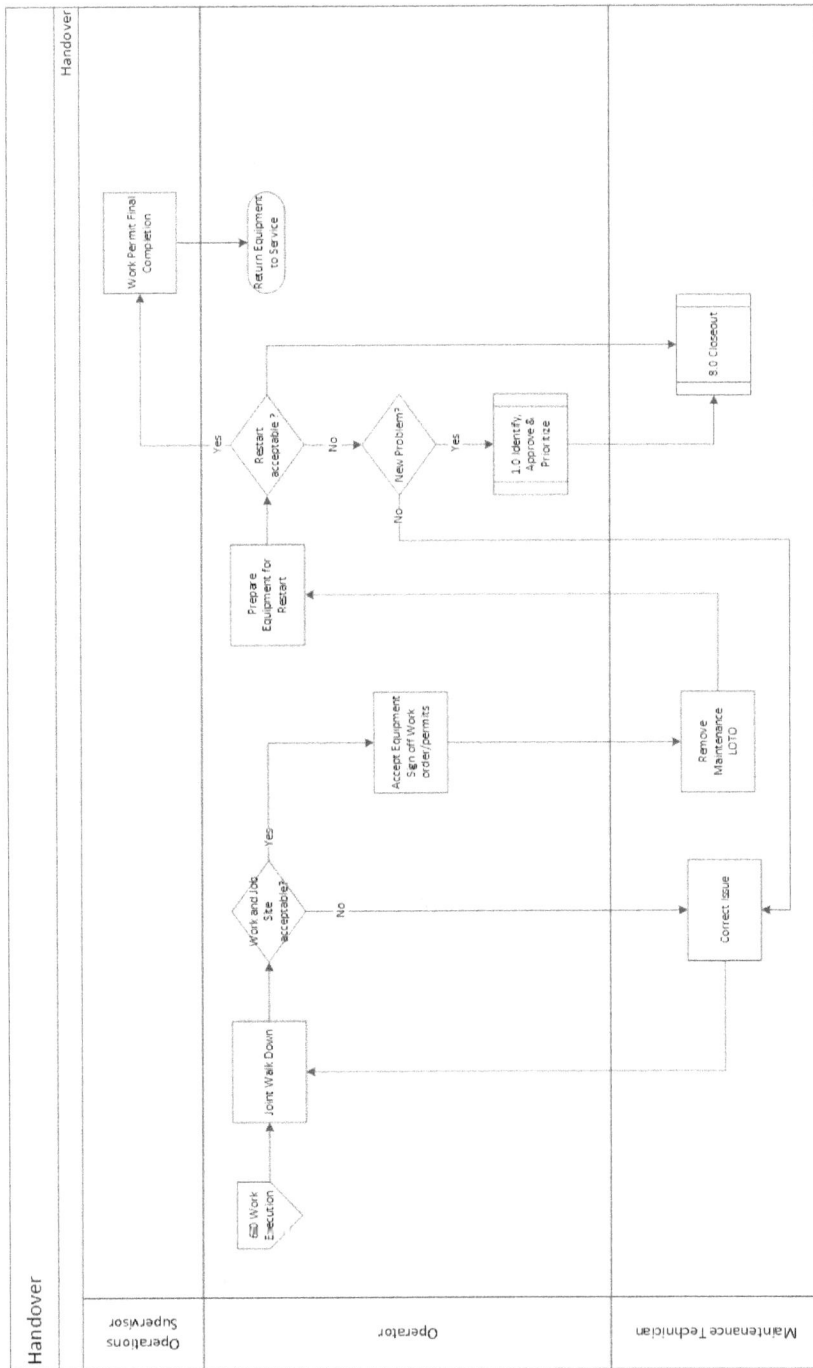

Figure 6-26: Handover workflow

It Ain't Over Till the Paperwork's Done
After the work is executed, and all the activities are complete, the maintenance technician completes the final steps:

16. The maintenance technician completes the work order.
17. The job is done.

Completing the paperwork has traditionally meant that the technicians would indicate how much time the job took, and include a record of any parts and material that might have been used on the assignment. Many organizations rightfully require their technicians to document any other aspects that need notation, and to sign the work order.

Ultimately, here is what we need to see, or better said, need to infer from the notes made on the work order:

- How the technician found it (the asset)

- What they did

- How they left it

Some technicians can complete this requirement in as little as three sentences or paragraphs. Still others will print out a dissertation. A few might start with "It was a dark and stormy night... " and end with " ...and they lived happily ever after."

It doesn't really matter. What does matter is that in two or three years, we can print out that work order and determine very quickly the story that the work order is telling us. We want to be able to discern how the technician found the asset, what they did, and how they left it.

It is very hard—no, make that impossible—to find the root cause to a problem with two years' worth of:

- Motor broke

- Motor fixed

That just won't get the job done.

With exceptional effort at identifying issues all the way through scheduling and executing the repair we can avoid the functional failure of any and all components.

By identifying the information we must have at the close of a work order, we are better able to analyze what our equipment is experiencing and what we can do to provide better reliability and maintenance.

Analyze This File

This is one aspect of maintenance work management that I'm afraid to say we just don't perform well and it may be because we don't know what to do or how to do it. There is even a probability that we don't even know who is supposed to do it: analyze the data, that is.

Figure 6-27 shows where *analyze* falls on the P-F curve.

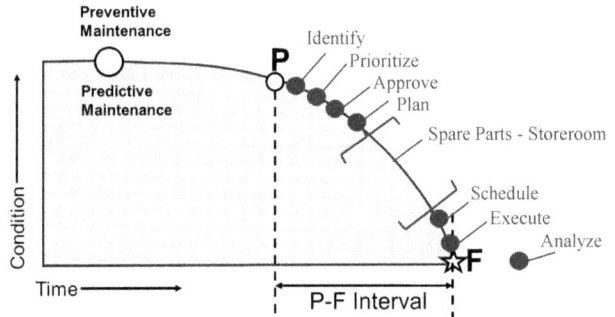

Figure 6-27:
Analyze nodule
on the P-F curve

You have, no doubt, noticed that the analyze nodule is on the curve after the point of functional failure. This was done purposefully to indicate that this particular work is done on-line, meaning the equipment is running, and this activity is accomplished after the component is returned back on the curve (through repair or replacement).

What does analyze mean, exactly, and who performs this analysis? There are actually two separate answers to these questions.

The Planner as an Analyst

Tho planner is the Alpha and the Omega of the work order system. No work orders are released that don't come from the planner (scheduler) and none are entirely complete that aren't finished up by the planner. Beginning to end.

As such, the planner has the expressed responsibility to review the incoming, completed work order and assess what the result was of the completed maintenance activity. In this way, the planner has a finger on the pulse of all the corrective work, and the resulting information:

- How it was found

- What was done

- How it was left

Likewise, when the planner is given a freshly minted work order, meaning a work request that was just approved, the planner can answer a very simple question. "Have I seen this before?" The planner is required to look back in the maintenance history to see if the work has ever been done before.

Here's an example. If a 10 Hp motor is failing, and the operator writes up concerns later verified by maintenance, then the work request would become a work order upon approval. The planner might look back in the equipment history and note that this particular motor has been changed three times in nine months. That is a trend. That is a concern. That needs to be analyzed to find out why.

There is a very good probability that replacing the motor for a fourth time is not going to finally fix the reasons that the motor is failing in the first place.

Yet we do this same type of thing every day. In fact, it might be that this sort of repeat work happens thousands of times a day throughout the manufacturing and service world.

Even Figure 6-5 points to the absolute need for analysis of "what's going on" with our PM program.

The planner is in the perfect spot on the organizational chart, and positioned perfectly to perform the analysis of the data to determine trends and concerns. Remember, one data point does not make a trend.

What tools do planners have at their disposal? What should get analyzed?

This is a quick list of the types of data that might be analyzed. At the end of this list, I've left room for you to add what you feel should be included.

- Components that have completed PM tasks, yet fail in service prior to the next scheduled PM

- Repeat failures (maintenance cause codes)

- Completed work orders that exceed estimated times

- Completed work orders that exceed cost (labor and parts)

- Asset breakdowns, total and areas where occurring (create what's called a *heat map*)

- Negative work order survey results (operators and maintenance)

- Job kits that are missing parts

- Asset availability denied or cut short

Add your thoughts here:

_____

_____

_____

_____

The planner uses any number of tools to complete this analysis of data. The purpose of the tool is to convert data into information. Some of those tools might include these more popular ones:

- Pareto chart

- Run chart

- Scatter diagram

- Day-by-the-hour chart

As these are usually well known, we won't go into a major discussion on them, but we should spend a short amount of time to discover some uses.

The Pareto chart has led to what many call the 80-20 rule. It does seem typical, in almost everything, that 80% of the issues stem from 20% of the sources.

For examples:

- 80% of my Human Resource problems come from 20% of my employees

- 80% of my budget is spent on 20% of my bills

- 80% of my downtime comes from 20% of my components

It is almost eerie how often this turns out to be true. Figure 6-28 is a very simple example.

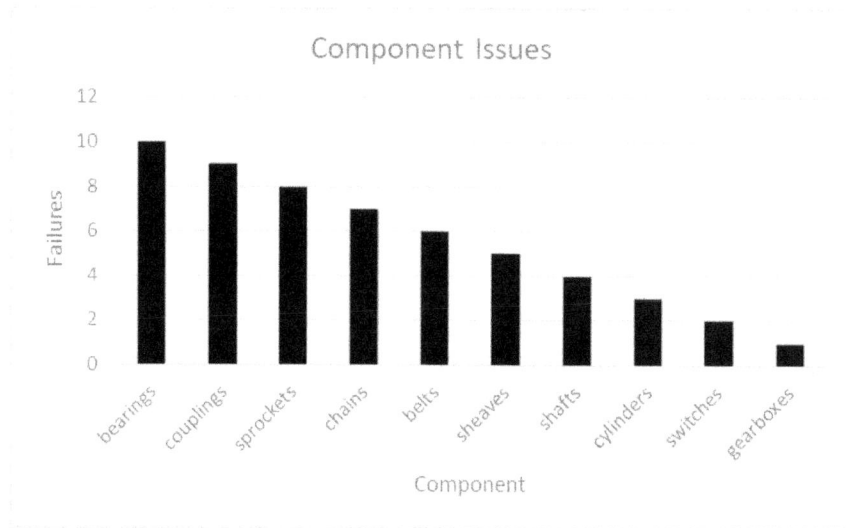

Figure 6-28: Sample Pareto chart

Figure 6-28 was purposefully made simple to show that 20% of the components were causing 80% of the problem. Of course, it doesn't always work out this cleanly, but it is helpful to consider what the data is telling you.

A run chart is a very powerful and thus, a useful tool. This particular product can tell me what my performance is, and how it has been trending historically. I'm going to offer an example of a run chart that considers a maintenance issue. Table 6-6 reflects work orders not correctly filled out by technicians by the technician's name.

Table 6-6: Work orders that are incomplete

Technician	W.O. Count	% of their total
Bill	7	15%
Fred	6	22%
Nancy	2	5%
Steve	0	0%
Kimberly	5	8%
Austin	5	12%
Rick	3	5%

From Table 6-6, it is clear that Bill has the highest raw work order count for incorrect work orders, but Fred has the highest percentage of work orders that are incorrect based on their number of completed work orders. Give some thought also to Austin, who only has five work orders that are incomplete, yet they account for 12% of his total. These are concerning numbers.

Figure 6-29 is a good representation of a scatter diagram indicating that we seem to have an absenteeism issue revolving around days 5-7 of a normal week (in our case, Friday-Sunday

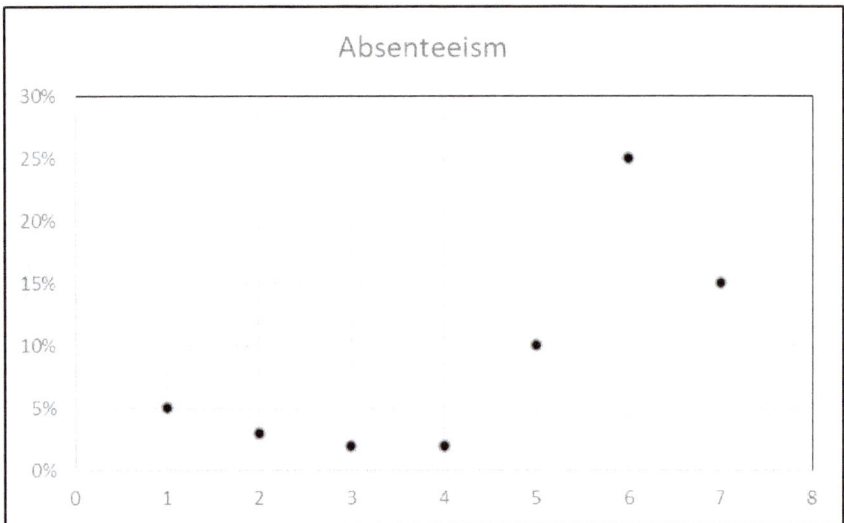

Figure 6-29: Scatter diagram example

A day-by-the-hour chart is perfect for collecting information in *real time* and determining any issues that affect normal operation.

These could be either good issues or bad issues.

For example, you might expect better interaction between maintenance technicians and the storeroom right after you held a meeting with all your maintenance crews on this very subject. You could measure that interaction before and after the team meetings.

Not only do the planners provide a valuable service in their roles as data analysts, but the reliability engineer does as well.

The Reliability Engineer as an Analyst

There really are only a few points to make on this particular topic. The reliability engineer has to be engaged in increasing MTBF and decreasing MTTR. Take into account all the discussions we've had over the pages of this workbook. This is their primary role.

For the probability calculations we performed in the scheduling section of this chapter, all the way back to the design of the equipment and the standardization of parts and the Bill of Materials, the reliability engineer needs to track and trend the performance of the asset.

Simple calculations and experience can be used to perform these small pockets of analysis-turned-action. Tracking the data, and demonstrating success (or failure) through charts and diagrams will help the reliability engineer make a compelling case for improvement for the operations of the plant.

The overall analysis process might very well look like Figure 6-30 on the following page.

Once the analysis is done and all the paperwork is corrected, and everything is complete, the job plan is ready to file away and will be ready for another day.

File

Figure 6-31 is the last step in our P-F Curve review. It indicates the time where the job plan is filed away for future use.

You may recall an example I provided many pages ago where I told the anecdotal story of being in an underground mine with the planner who was preparing job plans for jobs that had not even surfaced yet (pun intended). That planner was starting the process right there at the *file* node.

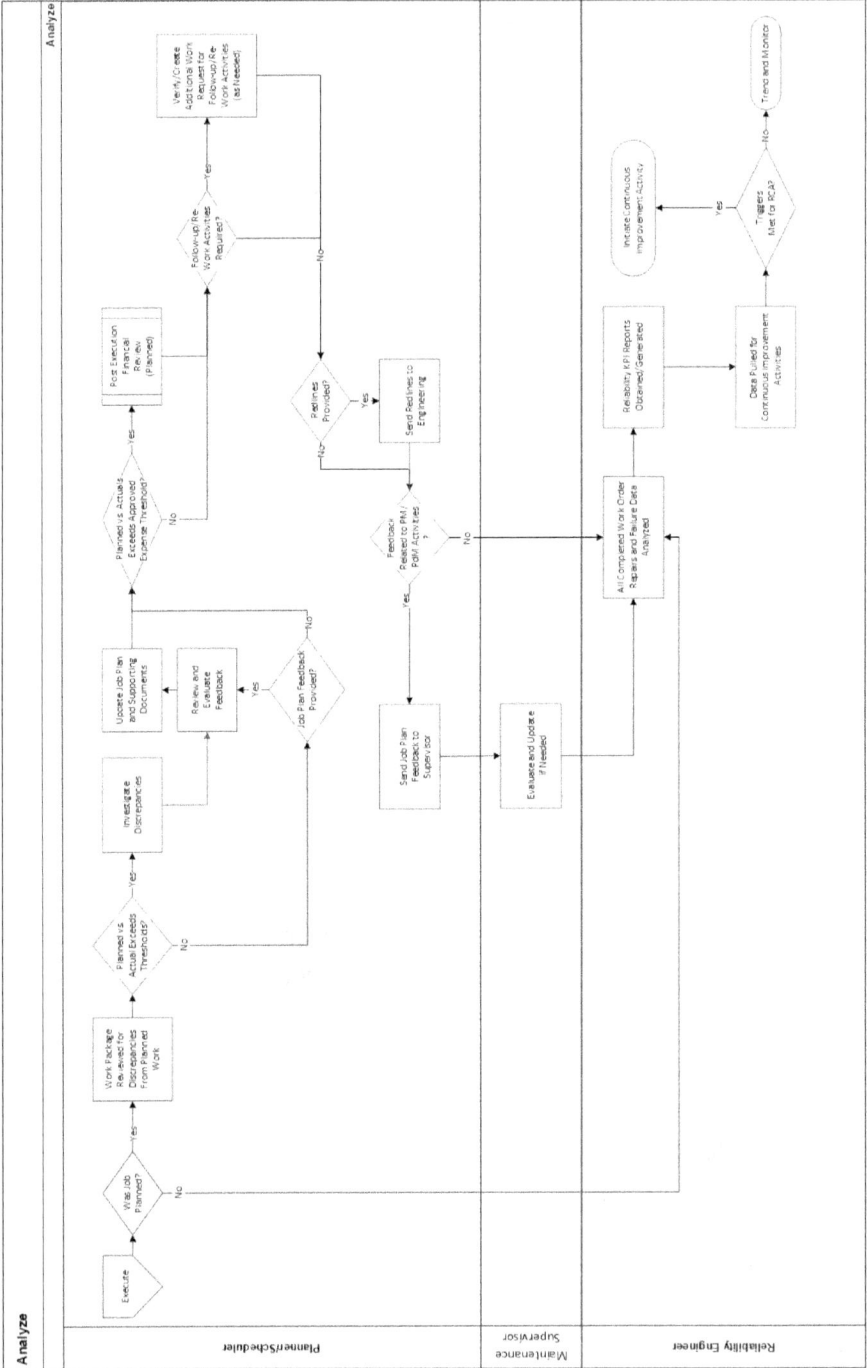

Figure 6-30: Analyze

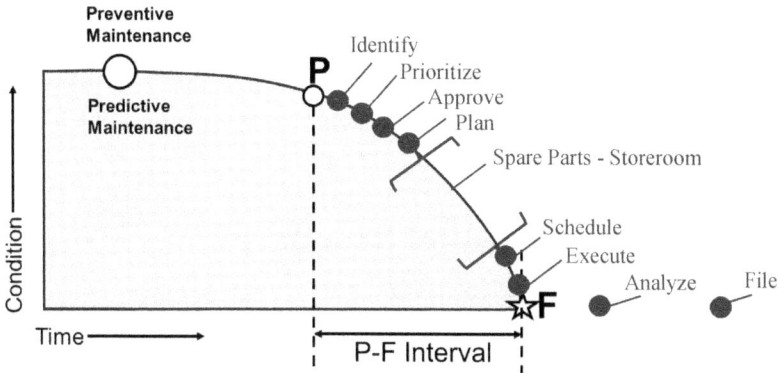

Figure 6-31: File nodule on the P-F Curve

The most important aspect of this particular point is to know how and where to label and store completed job plans for use at another time in the future. How do we know that we'll need to reference an old job plan and reuse it? You can bet that the new component we install will have its own P-F curve.

This may sound simple, but it is an amazingly complex process to devise a file nomenclature protocol and a filing scheme. There are so many Computerized Maintenance Management Systems, and within those, so many facets to understand and processes to adhere to. This can be a very complex issue. As such, it is very likely that an organization, maybe yours, has decided to create cheat-sheets to help employees understand, or perhaps they print out all the documents and file a hard copy for easier retrieval.

This particular element needs all the cunning and cleverness that can be leveraged from the maintenance and the Information Technology (IT) department. It may even require a small team meeting to determine what protocol is permissible and also how documents can be searched and located from other sister plants.

Some companies choose to file their job plans and maintenance history under the asset number. SAP users are likely to try a function-location process. Still others will simply try to remember a work order number. This is the hardest manner in which to accomplish this task.

It's best to figure out how to file the information before you begin to collect the information.

There is no doubt that you will create a process that works very well for your location. Based on what you've read throughout this text, and the exercises you took part in, you must have a solid sense of what is good and proper discipline when it comes to maintenance work management.

Chapter Summary

Congratulations! You have made it through this workbook and I'm certain you've found some valuable information to help with your reliability program and efforts. Chapter 6 was an especially engaging chapter as it speaks to the heart of what we do day-in-and-day-out in our maintenance operations.

Just to recap: in this chapter we used the P-F curve to pace our discussions through the heart of the maintenance work management processes. Specifically, we engaged in discussions centering on:

- Preventive and Predictive Maintenance
- Identifying work
- Prioritizing work
- Approving work

- Planning work
- Scheduling work
- Executing work
- Analyzing the work
- Filing the work for future use

Our developing philosophy:

We will organize and develop our maintenance processes to maintain the inherent reliability of the assets, staying primarily back on the P-F curve as shown on the following page.

While maintenance is working to maintain assets and components back on the P-F curve, we will also coordinate closely with production to ensure that our partners in operations are closely monitoring equipment and component performance and reporting those issues through the work order request process.

Maintenance and production will work together to create a RIME (Ranking Index of Maintenance Expenditures) Chart to determine the proper priority for work. A sample RIME Chart is

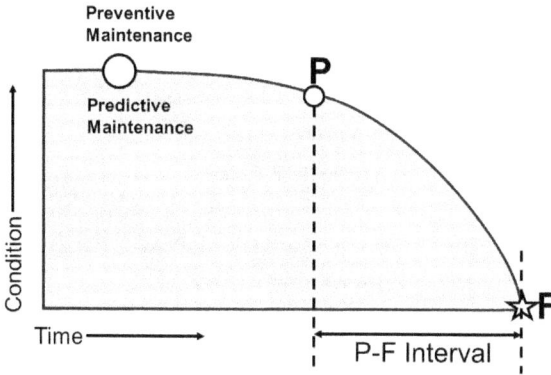

shown here:

Rank		10	7	6	5	2	1
Rank	Equipment	Breakdown Critical Safety	PMs	Corrective	Service	Routine	Facilities and Grounds
10	Main Power Air Compressor IT Network Furnace	100	70	60	50	20	10
9	Main Water Cutting Table 6 Line 1 Line 2	90	63	54	45	18	9
8	Cutting Table 8 Line 8 Line 00	80	56	48	40	16	8
7	Line 3 Line 7 TTUM 2000 STM	70	49	42	35	14	7
6	BND 17 Cutting Table 4/5 MUTT 4000 Hawkeye	60	42	36	30	12	6
5	Batch Saw Filler TFB Bender Sling Crane	50	35	30	25	10	5
4	Pattern Area Chopper Spider	40	28	24	20	8	4
3	Forklift Sweeper Docks	30	21	18	15	10	3
2	Trash compctr Rec compctr	20	14	12	10	4	2
1	HVAC	10	7	6	5	2	1

Priority 1
Priority 2
Priority 3

Once work is correctly identified and prioritized by the request-or, the request will be considered for approval within twenty-four hours. The approving authority will be the production leadership for any work request that requires that a production asset be shut down or taken out of service to complete the work. Otherwise the maintenance leadership will be the approving authority.

Work requests not fully and/or correctly filled out will be returned to the requestor for correction.

The requestor will be notified after the approval meeting as to the results of the review.

The maintenance planner will begin work preparation on all approved work orders. The maintenance planner will use a job planning checklist to lay out the required steps and material for a complete maintenance job. The planning checklist will include information such as:

- Work Order number
- Asset number
- Priority number
- Tools needed
- Equipment needed
- Material needed
- Parts needed
- Major procedural steps to complete the job
- Safety precautions
    - Energy isolation
    - Electronic interlocks
- Other outstanding work orders for the same asset
- Estimated time to complete the work order
- Skills required and number of people required for the work order

The planner will complete the job plan using the job planning checklist. It is during this planning phase for a maintenance assignment that the spare parts requirements are discovered. The planner will complete a listing of required spare parts and communicate this need via a pick-list to the storeroom.

The storeroom will endeavor to have the right part, at the right time and in the right quantity.

Once parts are verified to be available, the maintenance job will be ready to schedule. The maintenance planner/scheduler will have a completed one-week schedule ready each Thursday afternoon for the following week's work. Additionally, the scheduler will have a tentative schedule drafted up for two weeks out.

Our production partners are required to provide windows of opportunity for maintenance work, and the maintenance department will be ready and capable of making those windows to complete the work in the order of its priority.

In an effort to maximize the usefulness from each component, the maintenance work will be scheduled for, and will be executed as near the end of the P-F curve as possible, as indicated here:

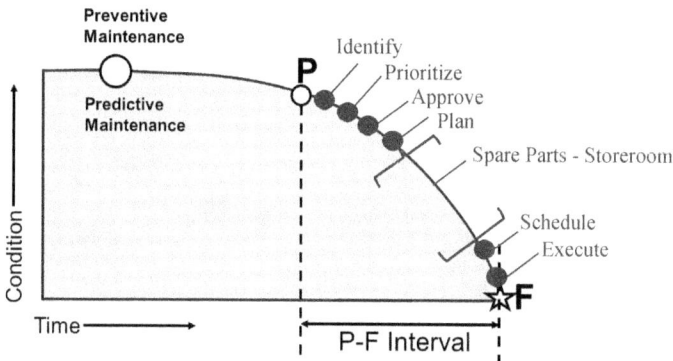

In all instances where the asset exchanges ownership before and after the planned maintenance job, both maintenance and operations will execute a responsible hand-off using an agreed upon turn-over and hand-over process.

After the execution of planned work, the maintenance department and reliability personnel will analyze the information gathered from the job and around the circumstances to effectively evaluate if the job that was planned was the actual job that was executed, or if the scope of work changed in some way. Additionally, the job will be analyzed to ensure the root concern was corrected.

Maintenance work will also be analyzed to determine if any trends exist regarding equipment or performance issues and to provide any conclusions that might be useful in the future.

The job will be filed away after all the close out administrative tasks are complete. The job will be filed in such a way that it can be retrieved for use in the future for that component or asset relevant in the job itself, or for consideration in any similar work in the future.

The End

Well, you've done it. You made it all the way through this workbook. If you've been working along with the exercises and chapter summaries, you have developed a very comprehensive maintenance and reliability strategy for you and your organization.

If you have not done so already, take time to compile your chapter summary thoughts by completing the Master Strategy form at the end of this workbook.

I hope you have found this book to be entertaining and informative. We need to start the debate and begin moving from *ideas to action*.

SUMMARY

In a way, I've been writing this book my whole professional life. I've always thought of myself as a student of reliability and maintenance, and still do. I don't think we ever stop learning. Or, at least we shouldn't.

I've always been focused on a *workbook*, believing that a workbook was the mode I felt would best get the point across that I was trying to make. I explained to others that I wanted a body of work that would reflect the same kinds of things that you and I might actually discuss if we were sitting across the table from one another.

In fact, many of the graphs and charts that are included in this text are the same pictures and drawings I would sketch out if we were sitting at your desk or a table somewhere. I actually use these tools weekly to make a point *stick*.

I want you to consider what you've read over these pages and how they might apply to your work life. If I've said something controversial, and touched a nerve, all the better. We don't need a revolution in maintenance and reliability, but we do need an evolution; as in, it is time to evolve.

Debate is good; finger pointing is bad. Let's not be afraid to have the tough and frank discussions we need to have. I hope some of this workbook will help to formulate the debate where you work.

There really is something you need to consider. I mentioned this in one of the opening chapters. If it isn't *you* who is going to lead this evolution, then who is it? If the time is not *now*, then when is it?

I think it is you, and I think the time is now.

The last section will be the continuous and cumulative documentation of the strategy you have been developing this entire time. Please refer back to the chapter endings and enter your notes once more to generate this decisive document.

Please use it to present a professional and well prepared argument.

Good luck!

The Master Strategy:
Our Maintenance Strategy and Philosophy

The developing philosophy:

Working in partnership, in a metered and measured manner, we intend to deliver world-class reliability efforts, meaning _____
_____
_____

We also want to attain and sustain the highest level of main-tainability, meaning _____
_____

Our aim is to obtain the highest sustainable level of uptime, as measured by _____.

We are certain that an optimized approach to effective and effi-cient work, demonstrated as, _____
_____

will result in higher output and an increase in team morale, evi-denced by _____
_____.

Our reliability belief is (originally recorded as a vision in Table 1-2.) _____
_____
_____

As such, we have expectations to deliver on these three specif-ic points:

1. _____;
   measured by _____;
   resulting in _____.

2. _____;
   measured by _____;
   resulting in _____.

3. _____;
   measured by _____;
   resulting in _____.

Furthermore, we believe that real ownership is derived by connecting responsibility with authority. Maintenance, production, engineering and purchasing will work together to ensure that those associates assigned the responsibility and accountability for a machine's reliability, or a process, will have a clearly defined level of authority. Additionally, the path to reach a higher level of authority will be unambiguously stated.

Improvement does not come from reliability increases alone. Additionally, we will work together to address waste in the area of maintenance, specifically:

Transportation: _____

Inventory: _____

Motion:_____

Waiting:_____

OverProduction:_____

Over Processing:_____

Defects:_____

And, we will work as a team to minimize or eliminate areas that can contribute to undesired variation in the service and products of the reliability efforts. Variations such as:

- Ambiguous task instructions

- Lack of specificity in PM steps

- Open ended maintenance time allotment

- Storage conditions of storeroom items

The goal of our reliability strategy is to decrease asset downtime and, by doing so, increase equipment availability for our production partners. Our four prong strategic plan is to:

1. _____
2. _____
3. _____
4. _____

Our clear and focused vision is:

_____
_____
_____
_____

And our mission is direct:

_____
_____
_____
_____

We have three concrete reliability goals. Those goals are listed here, as well as the means intended to reach those goals, and the consequences expected as a result.

Goal	Means	Consequences

Goal	Means	Consequences

Goal	Means	Consequences
_____	_____	_____
	_____	_____
	_____	_____

With this vision, mission, and a strategy that supports our intended goals, we are confident that we can reduce the maintenance budget each year. In a controlled manner, working with the partners we value, we will reduce the maintenance spending by 3% per year.

Objective evidence of our progress will be tracked by the following metrics and KPIs:

_____

_____

_____

_____

The compelling argument to improve our current reliability efforts is a powerful message. The organization will benefit in a fiscal sense, meaning: ($$$)_____

_____. And the value to our workforce will be evidenced by: (WIIFM)_____

_____

As we work together along this path of continuous improvement, we will communicate the progress of the individual activities by: _____

_____

_____

_____

Of specific value is the relationship between maintenance and operations, and maintenance and the Environmental, Health, and Safety divisions. Our work will be to improve reliability while meeting the safety and environmental goals of:

_____

_____

_____

_____

All this we will do while complying with the most basic communication needs as identified in the following diagram:

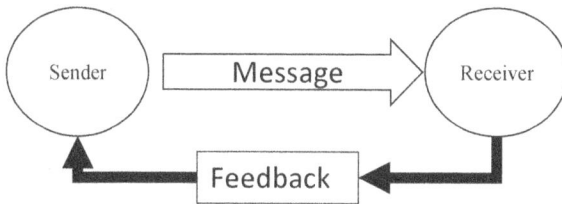

We will provide concise and complete communication as it relates to the reliability vision and anticipated deliverables. We will always seek, and be open to feedback in an effort to engage the entire workforce and give everyone a 'seat at the table' when discussing equipment reliability.

The maintenance and reliability organization will strive to establish a sense of process ownership within our associates. We will do this by marrying responsibility with authority. Our desire is for commitment over compliance, believing that ownership is the catalyst for commitment.

The processes that we will own together with our partners are:

- Identify
- Approve
- Prioritize
- Plan
- Schedule

- Turnover
- Execute
- Return to service
- Closeout
- Evaluate

The storeroom processes that we will own are:

- ABC Classification Review
- Adjust ABC Classification Model
- Add to Stock
- Bench Stock
- Critical Spare Parts Algorithm
- Critical Spares Review
- Cycle Counting Criteria
- Cycle Counting
- Data Scrubbing
- Disposal of Scrap
- Document Stockout
- Emergency Procurement
- Issuing
- Item Substitution
- Kitting
- Obsolescence Criteria
- Obsolescence
- Optimizing Stock Levels
- Ordering
- Preservation Program
- Receiving
- Repair or Replace
- Return to Stock
- Return to Supplier
- Salvage Value
- Special Tools
- Standardization
- VMI/Consignment

Each of these processes will be formally documented in process guides. All process guides will include the following sections:

- Business and Process Description
- Applicability
- Business Process Overview
    - Inputs
    - Outputs
    - Touchpoints
    - Key design decisions, requirements, and expectations

■ Business Process Workflow

■ RACI chart

■ Activity detail

Formally documenting these reliability processes will further enhance our ability to properly staff the efforts of the maintenance department by aligning the actual needs of the process with the roles and responsibilities of the associate.

From the processes, we will garner a sense of clearly defined expectations, leading to the creation of Key Performance Indicators that are in line with corporate, plant, and reliability goals. The KPI and goal alignment will be created using the balanced score card method, a sample of which is shown.

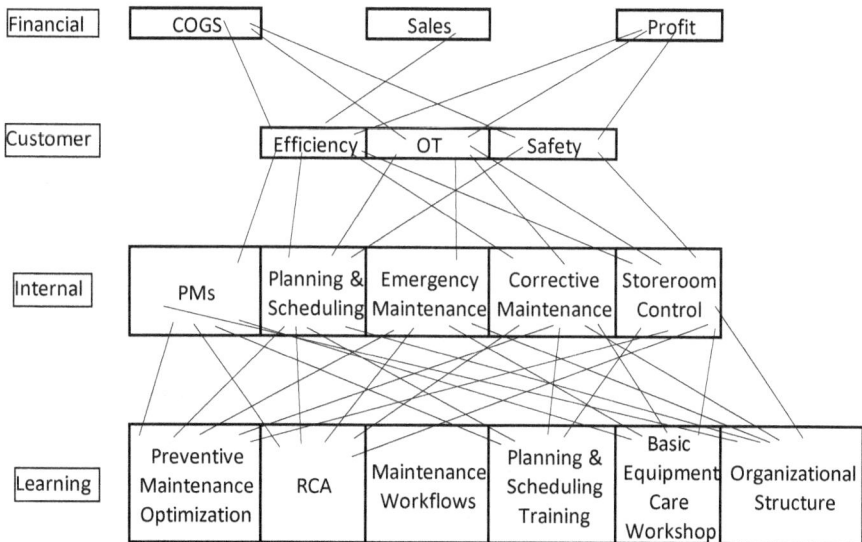

The net gain from this squaring of human resources to formally established processes, as well as expectations to KPIs, will be a maintenance organization that is a value added contributor to overall culture change.

We will aggressively attack problems by developing an exceptional problem statement technique. Our problem solving statements will take the form of:

- Who discovered the issue (who)?

- How big is the issue (what)?

- Where was the issue (where)?

- What time frame was information collected in (when)?

- The issue resulted in what, exactly (resulting in)?

Our maintenance organization is committed to weeding out issues, unfavorable trending of key metrics, and equipment reliability deficits. As such, we will employ the Root Cause Analysis (RCA) tool(s) of:

_____

_____

_____

_____

The processes that we will adhere to in order to maintain good order and discipline to high performing maintenance will be controlled through the process of Management Of Change (MOC). As part of the MOC program, any change that is not a like-for-like one will be vetted for technical feasibility and will be well communicated. This includes any change to rules and regulations.

Where possible and when applicable, we will research and cite rules, regulations, and suggestions for maintenance tasks to demonstrate the actual need, along with the actual process. This is meant to contribute to the ownership practice we will enhance in maintenance. As previously stated, these sources will be managed and controlled.

We further pledge to enhance, or create, a Process Safety Management (PSM) program to guarantee employee engagement and control of all safety and process requirements. This will be a leading indicator of our personal safety program.

Our continued passion for reliability going forward has to be rooted in the past. We know that an asset has to be designed, built, and installed in a manner that provides the greatest level of inherent reliability.

As this is the case, we will contribute manpower and knowledge to those making equipment decisions to offer the best technical advice on *D—design,* for maintainability and reliability, as well as *B—build,* with robust components that are already proving their worth in our environment, and *I—install,* to ensure that our equipment operates in a space that is conducive to a long and reliable life.

In addition to other metrics and Key Performance Indicators, we will employ classic Overall Equipment Effectiveness to measure the combined value of Availability, Performance, and Quality both before and after an asset modification. We will use a form similar to this one to provide the measures:

Machine:		Date:		OEE %	
Product:		Shift:			
AVAILABILITY					
1. Gross Available Time:	If 8-hour shift=480 min., if 12-hour shift=720 min				Minutes
2. Planned Downtime:	Time for meetings, cleaning, breaks, PM, and other scheduled events				Minutes
3. Run Time:	Gross Available Time – Planned Downtime	#1 - #2 =			Minutes
4. Downtime Losses:	a. Breakdown minutes _____ b. Changeover minutes _____ c. Minor Stoppages _____	a + b + c =			Minutes
5. Operating Time:	Run Time – Downtime Losses	#3 - #4 =			Minutes
6. AVAILABILITY:	Operating Time÷ Run Time x 100	#5 ÷ #3 x 100 =			%
PERFORMANCE					
7. Total output during operating time:	Parts, yards, units run per shift: good & bad				Total Units
8. Theoretical Cycle Time (ideal):	Minutes per unit				Min./ Units
9. PERFORMANCE:	[(Theoretical Cycle Time x Total output) ? Operation Time]	[(#8 x #7) ÷ #5] x 100 =			%
QUALITY					
10. Rejects during operating time:	(machine related defects)				
11. QUALITY	(Total output - Rejects /Total Output. X100)	[(#7- #10)÷ #7] x 100 =			%
OEE					
12. OVERALL EQUIPMENT EFFECTIVENESS	(Availability x Performance x Quality x100)				%

Our maintainability and reliability strategies will work to reduce Mean Time To Repair and Mean Time Before Failures. We will win with an approach that provides the highest level of maintenance and technical guidance.

We will partner with other agencies to ensure that our advice and guidance is founded in good fiscal practice, always providing a reasonable cost with a benefit that enhances, improves or guarantees asset reliability.

Along with ensuring that we build equipment that is reliable at the start, and supporting the asset through its life cycle, we will create a maintenance organization that can simultaneously address preventive maintenance, corrective maintenance, and emergency maintenance.

To ensure that we are delivering on the reliability promises, we intend to develop, measure, and report on reliability specific KPIs and suggest measures that other departments can take to ensure their staff are also supporting asset reliability.

To ensure that the proper level of reliability and maintenance management is available along with a balance of the necessary leadership in these disciplines, we intend to develop and execute a clear path of development curriculum that would include such courses, but not limited to:

_____

_____

_____

_____

To further hone and sharpen our reliability leaders, we will enact a practice of annual 360° feedback to build a solid team, working for what Deming called a *constancy of purpose.*

We intend to develop the desired level of leadership and management skills and abilities we expect. We will work with Human Resources and other such agencies to make certain that we have an action strategy to match our strong desire to grow our leadership.

The department leader will find a ready and capable crew eager for solid leadership. An organizational structure will be established to address the department's ability to respond to preventive, corrective and emergency maintenance simultaneously. The structure will be accompanied with a list of the necessary staff to make the reliability process work. The staffing members identified will have an accompanying listing of roles, samples on the following page.

Planner/Scheduler
Planner
Scheduler
Facilitator
Maintain BOMs
CMMS gatekeeper
PM/PdM overseer
Picklist generator
Data analyzer
Report generator
KPI/Metric generator

Storeroom Supervisor
Mentor
Assistant organizational structure developer
Assistant budgeter
Assistant stock determiner
Assistant report generator
Documenter
Facilitator
Assistant project manager
Supervisor
Stores Stock Committee Chairman

It is the expressed intention of the reliability group to develop a required skills listing and conduct an associated skills gap analysis. This analysis will be complete with an identifying legend and an indication of what the current state is, and what the path going forward will be in order to close the skills gap.

Our skills matrix will look similar to the following:

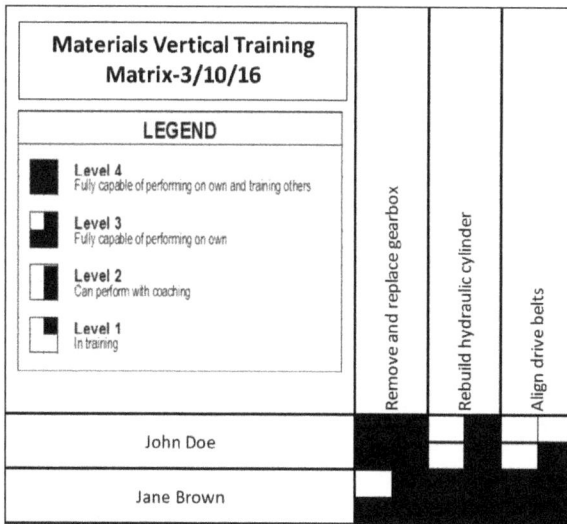

Materials Vertical Training Matrix-3/10/16

LEGEND

Level 4 — Fully capable of performing on own and training others
Level 3 — Fully capable of performing on own
Level 2 — Can perform with coaching
Level 1 — In training

Columns: Remove and replace gearbox / Rebuild hydraulic cylinder / Align drive belts

John Doe
Jane Brown

We plan to seek out and recruit sharp, excited associates in fields outside of maintenance to encourage them to consider joining this team. We believe that our production colleagues have as much interest in reliable equipment as anyone in the organization, and we intend them to have a chance to put their passion to work.

We will organize and develop our maintenance processes to maintain the inherent reliability of the assets, staying primarily back on the P-F curve as shown here:

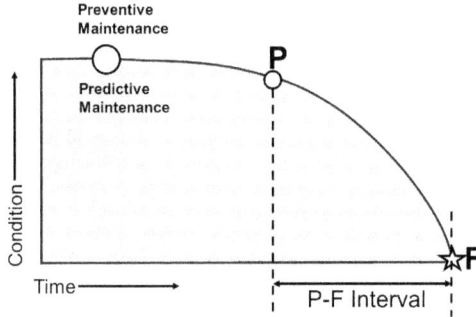

While maintenance is working to maintain assets and components back on the P-F curve, we will also coordinate closely with production to ensure that our partners in operations are closely monitoring equipment and component performance and reporting those issues through the work order request process.

Maintenance and production will work together to create a RIME (Ranking Index of Maintenance Expenditures) Chart to determine the proper priority for work. See RIME Chart shown here:

Rank		10	7	6	5	2	1
Rank	Equipment	Breakdown Critical Safety	PMs	Corrective	Service	Routine	Facilities and Grounds
10	Main Power Air Compressor IT Network Furnace	100	70	60	50	20	10
9	Main Water Cutting Table 6 Line 1 Line 2	90	63	54	45	18	9
8	Cutting Table 8 Line 8 Line 00	80	56	48	40	16	8
7	Line 3 Line 7 TTUM 2000 STM	70	49	42	35	14	7
6	BND 17 Cutting Table 4/5 MUTT 4000 Hawkeye	60	42	36	30	12	6
5	Batch Saw Filler TFB Bender Sling Crane	50	35	30	25	10	5
4	Pattern Area Chopper Spider	40	28	24	20	0	1
3	Forklift Sweeper Docks	30	21	18	15	10	3
2	Trash compctr Rec compctr	20	14	12	10	4	2
1	HVAC	10	7	6	5	2	1

Priority 1
Priority 2
Priority 3

Once work is correctly identified and prioritized by the requestor the request will be considered for approval within twenty-four hours. The approving authority will be the production leadership for any work request that requires that a production asset be shut down or taken out of service to complete the work. Otherwise the maintenance leadership will be the approving authority.

Work requests not fully and/or correctly filled out will be returned to the requestor for correction.

The requestor will be notified after the approval meeting as to the results of the review.

The maintenance planner will begin work preparation on all approved work orders. The maintenance planner will use a job planning checklist to lay out the required steps and material for a complete maintenance job. The planning checklist will include information such as:

- Work Order number

- Asset number

- Priority number

- Tools needed

- Equipment needed

- Material needed

- Parts needed

- Major procedural steps to complete the job

- Safety precautions

    - Energy isolation

    - Electronic interlocks

- Other outstanding work orders for the same asset

- Estimated time to complete the work order

- Skills required and number of people required for the work order

The planner will complete the job plan using the job planning checklist. It is during this planning phase for a maintenance assignment that the spare parts requirements are discovered. The planner will complete a listing of required spare parts and communicate this need via a pick-list to the storeroom.

The storeroom will endeavor to have the right part, at the right time and in the right quantity.

Once parts are verified to be available, the maintenance job will be ready to schedule. The maintenance planner/scheduler will have a completed one-week schedule ready each Thursday afternoon for the following week's work. Additionally, the scheduler will have a tentative schedule drafted up for two weeks out.

Our production partners are required to provide windows of opportunity for maintenance work, and the maintenance department will be ready and capable of making those windows to complete the work in the order of its priority.

In an effort to maximize the usefulness of each component, the maintenance work will be scheduled for, and will be executed as near the end of the P-F curve as possible, as indicated here:

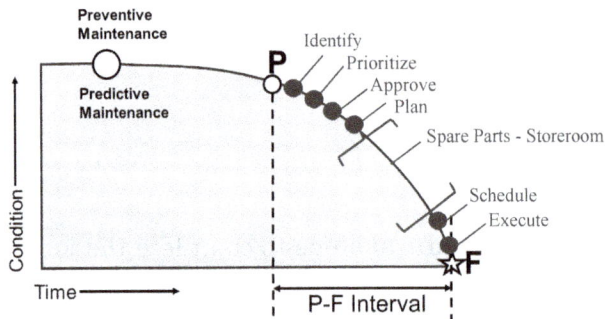

In all instances where the asset exchanges ownership before and after the planned maintenance job, both maintenance and operations will execute a responsible hand-off using an agreed upon turn-over and hand-over process.

After the execution of planned work, the maintenance department and reliability personnel will analyze the information gathered from the job and around the circumstances to effectively evaluate if the job that was planned was the actual job that was executed, or if the scope of work changed in some way. Additionally, the job will be analyzed to ensure the root concern was corrected.

Maintenance work will also be analyzed to determine if any trends exist regarding equipment or performance issues and to provide any conclusions that might be useful in the future.

The job will be filed away after all the close out administrative tasks are complete. The job will be filed in such a way that it can be retrieved for use in the future for that component or asset relevant in the job itself, or for consideration in any similar work in the future.

Through this carefully and thoughtfully created maintenance and reliability strategy, we can move together toward a partnership of engagement and constancy of purpose. With help and support, we can move *from ideas to action*.

REFERENCES

Abrams, R. (2005). *Business Plan in a Day*. Palo Alto, CA: The Planning Shop.

Aguayo R. (1991). *Dr. Deming, The American Who Taught the Japanese About Quality*. New York, NY: Fireside.

Austin, R. D. & Nolan, R. L. (Winter 2007). Bridging the gap between stewards and creators. *MIT SLOAN Management Review*, 29-36.

Belasco J. A. & Stayer R. C. (1993). *Flight of the Buffalo*. New York, NY: Time Warner Books.

Bormann, D. (2007, February 22). Turning out generals. *The Kansas City Star*, p. B1.

Bossidy, L. & Charan, R. (2002). *Execution: the Discipline of Getting Things Done*. New York: Crown Business.

Buckingham, M. (2005). *The One Thing You Need to Know, About Great Managing, Great Leading, and Sustained Individual Success*. UK Ltd: Simon and Schuster.

Collins, J. (2001). *Good to Great*. New York: HarperCollins Publishers, Inc.

Compton, W. D. (1997). *Engineering Management: Creating and Managing World-Class Operations*. Upper Saddle River, NJ: Prentice-Hall, Inc.

Darwin, C. (2017). *On the Origin of Species*. London, England: Macmillan Collector's Library.

Deming, W. E. (1986). *Out of the Crisis*. Cambridge, MA: Massachusetts Institute of Technology.

Gulati, R. (2nd Edition, 2013). *Maintenance and Reliability Best Practices*. New York, NY: Industrial Press, Inc.

Gundry, L. & LaMantia, L. (2001). *Breakthrough Teams for Breakneck Times*. Fort Lauderdale, FL: Dearborn Trade.

Harrington, H.J. (Dr.) (1991). *Business Process Improvement, The Breakthrough Strategy for Total Quality, Productivity, and Competitiveness*. New York, NY: McGraw-Hill.

Hill, N. (1960. *Think & Grow Rich*. New York, NY: Fawcett Crest.

Humphreys, J. (Fall, 2005). Developing the big picture. *MIT SLOAN Management Review*, 96.

Joel, M. (2009). *Six Pixels of Separation, Everyone is Connected. Connect Your Business to Everyone*. New York, NY: Hachette Book Group.

Kiyosaki, R. T. (1998). *Rich Dad Poor Dad, What the Rich Teach Their Kids About Money—That the Poor and Middle Class Do Not!* New York, NY: Warner Business.

Lee, R. D. (2018). *The Maintenance Insanity Cure, Practical Solutions to Improve Maintenance Work*. South Norwalk, CT: Industrial Press, Inc.

Liker, J. K. (2004). *The Toyota Way: 14 Management Principles From the World's Greatest Manufacturer*. New York, NY: McGraw-Hill.

Maxwell, J. C., (1999). *The 21 Indispensable Qualities of a Leader*. Nashville, Tennessee: Thomas Nelson Publishers.

Moore, R. (2004). *Making Common Sense Common Practice: Models for Manufacturing Excellence*. Linacre House, Jordan Hill, Oxford, UK: Elsevier.

Olve, N.G. & Sjöstrand, A. (2006). *Balanced Scorecard*. Southern Gate Chichester, West Sussex: Capstone Publishing, Ltd.

Paul, R. W. & Elder, L. (2002). *Critical Thinking: Tools for Taking Charge of Your Professional and Personal Life*. Upper Saddle River, NJ: Financial Times Prentice Hall.

Peters, T. J. & Waterman, Jr., R. H. (1982). *In Search of Excellence, Lessonfrom America's Best-Run Companies*. New York, NY: Warner Books.

Rath, T. & Conchie B. (2008). *Strengths Based Leadership*. New York, NY: Gallup Press.

Reynolds, J. (1994). *Out Front Leadership—Discovering, Developing, & Delivering Your Potential*. Austin, TX: Mott & Carlisle.

Slater, R. (1999). *Jack Welch and the GE Way*. New York, NY: McGraw-Hill.

Sutherland, J. (2014). *Scrum: The Art of Doing Twice the Work in Half the Time*. New York, NY: Crown Business Books.

Suzuki, T. (Edited by) (1994). *TPM in Process Industries*. New York, NY: Productivity Press.

Thoreau, H. D. *Walden*. Printed in the USA.

Walton, M. (1986). *The Deming Management Method*. New York, NY: The Berkley Publishing Group.

Verne, J. (2005). *Twenty Thousand Leagues Under the Sea*. New York, NY: Barnes & Nobel Classics.

Warfighting. (1994). *The U.S. Marine Corps Book of Strategy, Tactics for Managing Confrontation*. New York, NY: Double Day.

Wright, N.C. (Dr.). (2017). *The Death of Reliability, Is it Too Late to Resurrect the Last, True Competitive Advantage?* South Norwalk, CT: Industrial Press, Inc.

www.1000ventures.com/business_guide/crosscuttings/leadership_vs_mgmt.html, retrieved April 6, 2007.

webbjr.typepad.com/openloops/2005/03/mytwocentsle.html), retrieved April 6, 2007.

www.i-lead.com/articles/article002.html, retrieved April 6, 2007.
www.telusplanet.net/public/pdcoutts/leadership/LdrVsMngt.htm, retrieved April 6, 2007.

INDEX

www.ingramcontent.com/pod-product-compliance
Lightning Source LLC
Chambersburg PA
CBHW060752220326
41598CB00022B/2405